The Relic

A Marine's Memoir of the Korean War

By Richard A. Janca

♫♪

MARQUESAS PUBLISHING, LLC

Contributors & Acknowledgements:

Richard A. Janca – Author
Barbara Janca Rose – Introduction and Epilogue
Lisa M. Janca – Technical Editing and Formatting
Emily Janca – Factual Contribution
Dale Erickson – Factual Contribution- US Marine, Korean War Veteran

Marquesas Publishing, LLC – Publishing Preparation & Cover Design

Lynnita Brown – Editor – Founder of the "Korean War Educator"
Dawn Frederick Puzzitiello – Editor - daughter of the late Vincent
Charles Frederick, US Army, Korean War Veteran
Sue Roderick – Art Copyright Permissions – daughter of the late Rocco J.
Ferro, US Army, Korean War Veteran

ISBN: 978-0985461201

INTRODUCTION

This story is a marine's diary of the Korean War and the battle of Chosin Reservoir.

A story of courage, strong faith, and determination to lead others against incredible odds to become one of the "Chosin Few", Richard Janca, a young marine, is a hero who earned the Bronze Star and Purple Heart. A religious picture of the Boy Jesus was a relic he found amidst rubble and destruction. Finding the picture was an omen that gave him the faith and courage to survive. Richard was in many pivotal campaigns during the war. Richard carried the relic with him his entire life as a good luck memento.

We have provided a map of the Chosin Champaign, so the reader can follow the timeline and progression of the troops. Geographical information on this area is also noted for understanding the hardships the soldiers endured.

Chosin Campaign

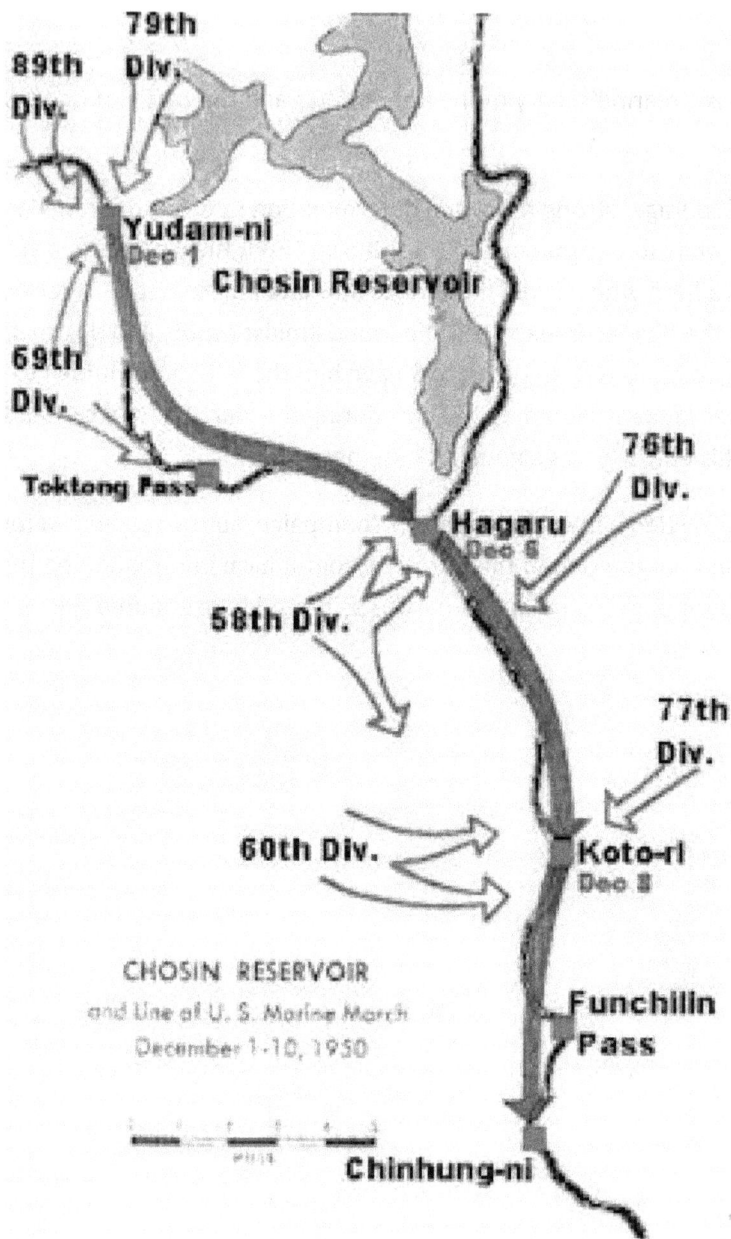

89th Div.

79th Div.

Yudam-ni
Dec 1

Chosin Reservoir

69th Div.

Toktong Pass

76th Div.

Hagaru
Dec 6

58th Div.

77th Div.

60th Div.

Koto-ri
Dec 8

CHOSIN RESERVOIR
and Line of U. S. Marine March
December 1-10, 1950

Funchilin Pass

Chinhung-ni

Background Information on Location, Terrain, Climate on Chosin Reservoir:

Chosin Reservoir is a man-made lake located in the northeast of the Korean peninsula. The name Chosin is the Japanese pronunciation of the Korean place name Changjin, and the name stuck due to the outdated Japanese maps used by UN forces. The battle's main focus was around the 78 miles (126 km) long road that connects Hungnam and Chosin Reservoir, which served as the only retreat route for the UN forces. Through these roads, Yudami-ni and Sinhung-ni, located at the west and east side of the reservoir respectively, are connected at Hagaru-ri. From there, the road passes through Koto-ri and eventually leads to the port of Hungnam. The area around the Chosin Reservoir was sparsely populated.

The battle was fought over some of the roughest terrain during some of the harshest winter weather conditions of the Korean War. The road was created by cutting through the hilly terrain of Korea, with steep climbs and drops. Dominant peaks, such as the Funchilin Pass and the Toktong Pass, overlook the entire length of the road. The road's quality was poor, and in some places it was reduced to a one lane gravel trail. On 14 November, a cold front from Siberia descended over the Chosin Reservoir, and the temperature plunged to as low as −35 °F (−37 °C). The cold weather was accompanied by frozen ground, creating considerable danger of frostbite casualties, icy roads, and weapon malfunctions.

Memoir Contents:

- **Background Information**   9
- **Training Camps**   12
  - **Boot Camp**   12
  - **Camp Lejeune**   12
  - **War Breaks Out**   13
- **Camp Pendleton**   15
- **Weekend Pass**   16
- **Trip to Korea**   18
- **Land of the Morning Calm**   20
- **Seoul City**   21
- **Moving with Dog**   25
  - **Mortar Barrage**   25
  - **Road to Uijongbu**   27
- **Operation Yo Yo**   31
- **North to Chosin**   34
  - **Schoolhouse Action**   37
  - **Road to Yudam-ni**   41
  - **Back to Hagaru**   46
  - **Yudam-ni Marines**   48
  - **Army Stragglers**   51
- **Breakout to Kotori**   53
  - **East Hill**   53
  - **Hell Fire Valley Again**   56
  - **Human Chain**   57
  - **Blessed Snow Cover**   58
  - **Bridge Repair**   60
  - **Tagged for Evacuation**   60
  - **Destination Pusan**   61
- **Masan: A City within a City**   63
  - **Beating the Odds**   63
  - **Tempus Fugit**   64
  - **Defective Ammo**   66
  - **The Chosen Few**   67
  - **Moving into Combat**   68
- **Pohang Guerrilla Hunt**   69
  - **Fast-rising Water**   71
  - **End of Vacation**   73

- **Decimated Army** 75
  - ○ **Holy Picture** 76
  - ○ **Tough Going** 76
  - ○ **Dignitary** 77
- **Combat Breeds Death** 79
  - ○ **Worthless Army Boots** 82
  - ○ **Chinese Casualties** 82
  - ○ **Soaked Through and Through** .. 83
  - ○ **Operation Mousetrap** 84
  - ○ **Caution around Civilians** 87
- **Time for a Break** 89
- **Back on the Line** 91
  - ○ **Savoring the Beauty** 91
  - ○ **Death at Any Moment** 92
  - ○ **Right Place, Right Time** 96
  - ○ **Souvenir Rifles** 99
  - ○ **On the Move Again** 99
  - ○ **Special Treats** 102
  - ○ **Vehicles and Horses** 103
  - ○ **Dear John** 104
  - ○ **Innovations** 105
  - ○ **Feeling the Heat** 106
  - ○ **Dropped from the Air** 107
- **Short-timer** 108
  - ○ **Special Assignment** 109
  - ○ **Crawling with Ants** 110
  - ○ **On the Rotation List** 111
  - ○ **Leaving my Unit** 114
  - ○ **Onboard the McKinley** 116
  - ○ **Pizza Pie** 119
  - ○ **Travel Misfortunes** 119
  - ○ **War Behind Me** 121
- **Post-Military** 124
- **Awards Received** 125
- **Epilogue** 126
- **Obituary - Richard Janca** 127
- **The Messenger** 128

The Relic – A Marine's Memoir of the Korean War
by Richard A. Janca

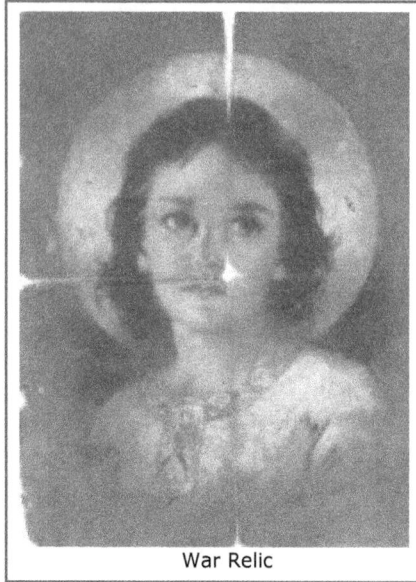

War Relic

KOREA
September 1950 - September 1951
W-2-7 1st Marine Division

San Diego, California. September, 1950

"It's early morning and it looks like we're in for another beautiful warm summer day. It's also a special day for me; in a few more hours I'll be embarking on the greatest adventure of my life. There is a war taking place in South Korea with a Marine Brigade already actively engaged since early August helping to stem the tide of a communist takeover. I've come a long way since I left home in January of 1948."

Background Information

I was born Richard Antoni Janca, the middle child of Joseph and Julia Zynczak Janca, on July 5, 1929 in Lackawanna, New York during the year of the great stock market crash. My father worked as a coalminer in Shenandoah, Pennsylvania after emigrating to the United States from Poland. Thereafter, he moved to Western New York upon hearing about the plentiful jobs in Lackawanna at Bethlehem Steel Plant. My mother was born in Lackawanna and was a war defense plant worker at Trico, in Buffalo, New York. My parents met and later married, then built a home at 100 Elkhart Street in Lackawanna, where all their children were born. I had an older brother, Edward, and a younger brother, Michael.

Lackawanna, known as the Steel City, got the nickname from a major steel-making corporation. Bethlehem Steel, the largest employer of labor in the area. I grew up with the Depression generation and all the hardships that came with it. Those were tough times, but they also taught one how to survive on the barest necessities in life. It was nothing compared to what Marine veterans of the Pacific War in my unit had endured. They not only had to deal with the enemy in combat, but also exposure to nature. The hot, infested jungle with its panoramic assortment of disease, plagued many of the veterans that were still around. Not long ago on a rifle range, I watched as a Pacific War veteran came down with a malaria attack. His canteen cup and hands shook so badly that he needed assistance to hold and drink from it.

I attended St. Michael's Catholic School and Washington Elementary School, and then I entered Lackawanna High School, from which I graduated in 1947. Like my father and brother before me, I applied for work at the local steel mill (Bethlehem Steel). I was hired immediately and was assigned as a laborer in the powerhouse. This would only be a temporary job until fall, when I would return to school to further my education and hopefully upon completion of study, find a better paying position. I almost didn't make it. One night while at work, I was assigned to a work detail, shoveling dirt out of a boiler. I can't remember what happened, except I woke up in the medical building lying on a cot. I asked, "What happened and how did I get here?" I was told I had collapsed while working and was brought to the hospital for medical treatment. I was overcome from gases that were still present in the boiler and was fortunate that I wasn't alone. The other workers saved me from what could have been a major disaster. This job had almost cost me my life, so I decided to enroll at Erie Community College.

In the fall, I was a full-time student attending Erie Community College, taking a course in mechanical technology. The school was located roughly ten miles from home. Since I had no wheels of my own, I had to commute to school by bus. I had a rather long school day, so I would on occasion, stop at Kogut's Tavern, a local gin mill and a regular hangout for most of my friends. It was shortly after the start of my second semester that I dropped in at the tavern for a drink and to chitchat with my friends. On this particular night I happened to arrive as a drinking party was taking place and I was invited to join them. There were six or seven in the group and all were hyped up about joining the service. In the group were recent Army and Navy veterans who wanted to try a tour of duty in the Marine Corps. With beer flowing freely and time slipping by, my friends convinced me it was time for me to leave home, do some traveling, and get to see some of the world. I agreed to join them the following morning at the local recruiting office. Needless to say, only three of us showed; the others either drank too much or had no intent to make a firm commitment when sober.

The next morning at the recruiting office, a Marine NCO extended a warm welcome, asking if we wanted to enlist. We came to enlist, so he filled out all the necessary paperwork and had us sign our names to a three-year enlistment. Now all that was left was a physical, scheduled for January 15, 1948, with departure to take place that same night if we were fit for duty. We were all accepted. Henry Martin Bodziak, a classmate of mine at Lackawanna High School and I, went to Parris Island for boot training, while John Pluta, my neighbor who lived three doors away and was an Army veteran, was issued orders to Quantico, Virginia. At a later date, I learned that some of my friends who originally chickened out in joining the Marines with us had eventually wound up entering other branches of the military.

Training Camps

Boot Camp

I remember my boot camp training days at Parris Island, South Carolina where discipline was taught in a punishing manner. I can still recall the day our Drill Instructor wouldn't excuse anyone to make a nature call. This resulted in an epidemic of pants wetting when a boot couldn't hold back any longer. Another time we ran the obstacle course and then double-timed all the way back to our barracks. The D.I. then asked if there was anyone that still had the energy to run some more. Three others and I remained in the ranks when the order was given to fall out. We were then instructed to run up and down the Company Street. After considerable time elapsed the D.I. came out and terminated our marathon. Once inside the barracks, our failure to admit fatigue resulted in demeaning punishment. We were ordered to get our buckets loaded with soap and water and then were told to scrub the barracks floor with a toothbrush. Discipline was a key to success and the Marines were good at it.

After graduating from Parris Island, I went to Camp Lejeune and attended Sea School. I applied to go to the Annapolis Naval Academy but the paperwork got lost and I was given different orders and assigned to the aircraft carrier USS Leyte CV32. I spent a two-year tour of duty aboard that aircraft carrier with the Sixth Fleet in the Atlantic Ocean, including one week aboard a destroyer during an amphibious training exercise against the island of Crete in the Mediterranean Sea. A guy named Harry Griswald and I became good friends while serving on the ship. At the end of our tour of duty on the Leyte, Harry and I received our transfer orders off the carrier on the same day. Our new destination: Camp Lejeune, North Carolina.

Camp Lejeune

On arrival at camp, Harry was assigned to the 2nd Marine Regiment. I was placed in the 6th Regiment, Weapons Company, 2nd Battalion. On reporting to my company, I was given a choice of 81 mm mortars, heavy machine guns or anti-tank platoon. Why I chose anti-tanks I

12

can't recall. I soon learned that this platoon handled 3.5 rocket launchers, flamethrowers, demolitions and later, light machine guns. Our platoon ran field exercises for reservist who came to camp for their annual two-week drill. In turn, we staged a lively show, attacking bunkers using flamethrowers and demolitions while our guests observed from bleachers on the sideline. Little did we realize while staging these exercises that a war was brewing and ready to erupt in South Korea. All these exercises were a training asset for all Marines involved in preparation for a real confrontation about to happen.

War Breaks Out

On June 26, 1950, we were informed at a special troop formation that North Korea had invaded South Korea during the night. There was no mention that America would be involved. My enlistment was to be up on January 14, 1951, but my "short-timer" status was shattered by Harry Truman's decree to extend my tour of duty one additional year. I had only had one ten-day leave since my enlistment, and that was time I spent at home upon completion of boot camp. I had accrued a sizeable chunk of leave time, so I applied for a thirty-day furlough. It was approved and my leave began on July 19, 1950. I came home from Camp Lejeune by train with a return ticket to use when my leave ended.

I had been home a few days enjoying myself when a special telegram arrived ordering me back to Camp Lejeune immediately. That telegram changed my leave plans. I decided to spend an extra day at home and then book air passage back to D.C., returning the rest of the way by rail. It was at the D.C. train depot that I bumped into my old shipmate Harry, whom I hadn't seen since our arrival at camp. He told me that he was newly married and was on his honeymoon having a great time when his telegram arrived, spoiling what time he had left of his leave. We rode back together, reminiscing about our good old days aboard ship.

Our arrival at Camp Lejeune had a few surprises in store for us. Harry Griswold's regiment had departed for California, with most of my regiment in tow. My battalion, I believe, was all that remained at Camp

Lejeune, along with some headquarters personnel. With his original company gone, Harry needed a home, so he was assigned to Weapons Company. It was just like old times being together again. I was placed on work detail, boxing equipment, and supplies on railroad cars for shipment to California. In the next few days, Marine replacements began to arrive from duty stations throughout the country. Rifle companies were reformed and the Sixth Marines received orders to relocate to Camp Pendleton, California.

Camp Pendleton

We made the trip to California by train. I remember stopping at El Paso, Texas. This was the only stop where we had a chance to get off the train to stretch and exercise. The temperature didn't help. It was just too hot. Shortly after our arrival at Pendleton, there was a change in regimental colors. We then became part of the Seventh Marine Regiment assigned to the 1st Marine Division. Our company and battalion remained the same.

It was a pleasant change of climate in California. At Camp Lejeune, the weather had been hot and humid, and more than one shower a day was normal to cool us off. What made it stickier and more unbearable was the lack of air conditioning. That kind of comfort just wasn't around in any barracks. At Camp Pendleton, it got hot during the day, with showers taken in an outdoor facility. The mountain air cooled off so much in the evening that we needed blankets to keep warm at night.

Our time at camp was spent training for duty overseas. Classes in Judo were overseen by John Ivers, a black belt holder. Demolition usage was held in the classroom and then actual testing was conducted in the hills. I can still recall a charge I assembled with my instructor observing. I tossed my explosive device and waited. Nothing happened. That's when my instructor told me to move my butt and retrieve it. I was in the process of rising off the ground when the instructor's hand pulled me back down to the ground. He said, "Wait a minute longer." Just then the charge exploded. The temporary delay saved me from what could have been a tragic accident, if I had gone to recover the charge when first told.

During our classroom sessions, the instructors went out of their way to highlight dangers involved when working with explosives, citing cases of injury or death resulting from careless handling of explosive material. After my near miss, I was in total agreement with their presentation.

Weekend Pass

During our short stay in sunny California, Harry and I got a weekend pass and we decided to make the most of it. We thumbed a ride to Los Angeles with our goal to see Hollywood. It was a Saturday night. We checked into a hotel with a tour of Hollywood planned for the next day. In the bar, we met two girls from Texas who were on a holiday away from home. One girl was married. The other one came along to keep her company and out of trouble. We had a pleasant evening.

The next morning after breakfast we set out to see Hollywood. Thumbing a ride, a car stopped and gave us a lift. The driver asked where we were headed. After making our destination known, he said, "I know just the place where you want to go." All the time we spent in his car, the driver filled us in on what was taking place in town. His communication skills appeared as if he was putting on an acting display for our benefit, until we arrived at a tavern that the driver said would make our day. That trip was a novel experience for me. It was one that I shall never forget.

Inside was a hot band holding a Jam session. The place was mobbed and Harry and I were having a great time. I got introduced to a beautiful girl, with jet-black hair. Her name was Irene. As the hours and day ebbed away, Irene got real talkative and told me that she came to Hollywood hoping to become an actress, but being of Polish ancestry, felt that that may very well hinder her acting ambition. Later that evening, she asked me if I would like to attend a late night house party at Al Jolson's place, the famous singer. The people in the group she was with were all going, and Harry and I were asked to tag along. It was a tough decision to turn down her offer, but we didn't have our own wheels and had to rely on being able to hitch a ride back to camp to make roll call in the morning. To this day, I've always wondered what I missed by not attending her party.

We never did see much of Hollywood. After leaving the tavern, we walked over to a gas station to ask directions back to camp. The owner asked what part of the States we were from: when I said Buffalo, New York, the owner perked up and said he was also from Buffalo. This led to a friendly exchange and we got to meet some of his family. Later, he gave us directions on how to find our way back to camp.

One day, a photo session was held for all members in our company. Anyone interested had an opportunity to purchase a keepsake company picture or a large individual color photo. I ended up with one of each and had them mailed to my parent's home. There was a night before our departure from camp that had Hoagy Carmichael down to entertain the troops, displaying his talent with a piano and songs he made famous.

Just before leaving the States, Harry came over and told me he was given a transfer to Dog Company. His prior experience in a rifle company, while with the 2nd Marines, was responsible for the transfer. Word came to saddle up and I returned to my section, not realizing that I would never see Harry again. A chaplain was in the area chatting with the troops. I noticed quite a few Marines had stopped to ask questions, some pouring out their inner feelings and asking for the Lord's protection, while others who had written letters now used the chaplain as the postman in handling their mail.

After two and a half years of duty, I was now a Corporal assigned to Weapons Company, 2nd Battalion, 7th Regiment, 1st Marine Division. My duty assignment was to act as guide for elements of my company due to arrive from Camp Pendleton sometime that morning. I directed them to their quarters aboard the U.S.S. Bayfield, APA-33, a Navy transport that was waiting for their arrival.

Trip to Korea

Finally, with the troop arrival, there was activity taking place on the pier as American Red Cross workers prepared to greet Marines, handing out sewing kits, missalettes, and copies of older magazines to anyone willing to accept them. While troops waited to go aboard ship, coffee and doughnuts were available to all that wanted to sample any of the freebies provided. A Marine band arrived just as troops started to go aboard ship. It was late in the afternoon with everyone on board. Our ship slowly started to leave port with the band still playing. "Harbor Lights" and "Goodnight Irene" echoed as Marines aboard ship joined in a final vocal rendition for a farewell to be remembered. The outline of San Diego slowly disappeared as daylight faded into darkness.

As mentioned earlier, I had spent a two-year tour of duty aboard an aircraft carrier with the Sixth Fleet in the Atlantic Ocean, including one week aboard a destroyer, during an amphibious training exercise against the island of Crete, in the Mediterranean Sea. I was now looking forward to adding a new chapter to my list of world travels. This would be my first Pacific Ocean crossing, with a promise of new adventures waiting. Time would tell. Our destination was somewhere in northern Japan. There was a rumor going around that had us training in Japan for a few weeks, and then we were to join the 1st Marine Division in Korea.

To occupy our time aboard ship, a training schedule became a necessity. Many newly-activated reservists were assigned to our units, some joining our ranks less than a month ago with no prior boot camp training. Daily drills were held for their benefit on weapon nomenclature, taking firearms apart and reassembling. Our main support weapons were flamethrowers, 3.5 rocket launchers and demolitions. Every member in our unit had a designated assignment. One Marine was to carry a rocket launcher with other team members hauling rocket shells. Flamethrowers, being heavy, were to be hauled aboard a company vehicle until needed. This also applied to demolition equipment. Each man's assignment had to be interchangeable within the team, in case a need arose.

With the schooling, there also had to be some time for fun and games. We couldn't have a boring ocean crossing with all work and no games. What could be more entertaining than having a Marine aboard display his talent in hypnotism? He had an ample supply of candidates volunteering to be hypnotized. The entertainment came as his subjects were put to sleep and then our fun began; it was hilarious watching grown men succumb to sleep and then obey his commands. It felt like old McDonald's Farm had taken over our ship, with subjects under his spell portraying animals with sounds and gestures identical to animals that his audience demanded. It wasn't a boring trip.

As we got closer to Japan, there was a change in weather conditions with typhoon turbulence causing pitching and rocking that resulted in many aboard becoming seasick. I was assigned a bunk in the bow area of the ship, where stormy seas were most evident. We had to ride out the storm and the best place for me I decided was in my bunk until the weather changed. Japan and Korea were the hardest hit by nature's fury during the height of this storm. On the morning of September 15, 1950, our P.A. system announced a successful landing of 1st Marine Division units at Inchon, Korea. This came as a surprise, since we were scheduled to arrive in Kobe, Japan that very day. Excitement spread through the ranks, giving rise to whether our plans would change as a result of this landing. A decision in change of orders was made upon docking. There were Marines waiting at Kobe for transportation to join the rest of the 1st Marine Division in Korea. It took thirty-six hours to square away our needs, and then we were back at sea.

On the morning of September 21, 1950, we arrived at Inchon with enemy targets ashore being shelled by a Navy task force that included cruisers and the battleship Missouri. I was on deck with other Marines scanning the shoreline, wondering what the future had in store for me. I wasn't the first or last in my family paying a visit to Korea. At the end of World War Two, my older brother spent a tour of duty with the army of occupation in Korea. Upon his return home, he told of the many problems experienced with natives who supported the communist movement. My younger brother, who enlisted in the U.S. Air Force, also got his chance for a tour of duty in Korea.

Land of the Morning Calm

It was time to disembark. Our sea bags with all unnecessary clothing and personal possessions were to be left aboard ship. Each Marine wore his combat uniform with appropriate equipment, plus leggings. On the pier, trucks were waiting to take us to our prearranged assembly area. With all the troops aboard, the trucks moved out. It was late in the afternoon when we finally got to our designated area. Maps were checked to verify our location. It turned out that we had overshot our destination and were now way out into no man's land. Our company officers held a discussion, deciding to march us back toward our designated battalion position. During our trek back, darkness took over. We were given orders to dig in for the night. As we dug our foxholes, word filtered down that Marines a short distance from our location had dug into a gravesite, their shovels making contact with bodies buried beneath the earth. This alerted everyone to use caution in his digging. I for one had no such experience.

During the night, as I watched from my foxhole, a firefight could be seen taking place quite a distance from where I sat. Tracers lit up the skyline while shell explosions echoed in the night. We were late for the original landing, missing all the fireworks. This was a preview of what we missed and what awaited us in the very near future.

The next day we relocated to a new position. It was still daylight when word came to dig in. Since it was early, it gave each pairing partner a chance to dig deeper foxholes. A shot rang out to my right; I dropped my shovel, grabbed my rifle, and waited to see what would occur next. Shortly, word came back that "Sergeant C", a Pacific war veteran, had shot himself in the foot while making a nature call. He was our first casualty, giving rise to questions whether it was accidental or intentional. Later that night, all hell broke loose, as machine gun and rifle fire opened up a few yards to my right. It lasted for a few minutes and ceased as quickly as it began. It turned out a Marine in a heavy machine gun position had crawled out of his foxhole to make a nature call when someone with an itchy trigger finger let loose. This Marine did manage to make it back safely to his foxhole, but there was a lesson for everyone to learn and remember. Anyone moving at night was fair game to others around him.

The following morning we moved out to what looked like open terrain. There, we formed a skirmish line moving through the ground in front of us, not knowing what to expect. There was no enemy activity. We came to a road late in the afternoon. Up until then, we carried our full packs and weapons. A short break took place and then we were told to drop our packs, but to be certain to carry our shovels. Our packs were to be placed aboard Jeeps and would be available to us later that night.

It was a warm day when we started, so even our field jackets were left behind. Whoever made this decision made a terrible mistake. We proceeded to climb a ridge, got to the top and stopped just before dark. That location was to be our home for the night without our field packs. The weather changed and the temperature dropped. The colder it got, the more I missed my pack with field jacket and sleeping bag. There was no way to keep warm. I suffered most of the night and was glad to see the sun rise. The cold weather prevented most Marines from getting any sleep during the night.

We recovered our packs and made our way toward the Han River. A great deal of activity could be heard and seen taking place on the other side. We came to a newly constructed pontoon bridge with traffic moving both ways. We finally got across, and that's when I saw the deadly price Marines were paying for real estate being liberated. In trucks coming back from the combat area were bodies of dead Marines piled up one on top of the other, just like stacked cordwood. We moved until dark and dug in for the night on the outskirts of the city of Seoul.

Activity taking place in the city was unreal. Planes were flying overhead dropping their bomb loads. The city was on fire with the sky reflecting a glow of destruction that was taking place below, with shock waves shaking the ground we occupied. This was the closest we came to actual combat since our landing.

The following morning, my section under Sgt. McBride was ordered to join Dog Company. Our mission was to enter Seoul from the north. The road we marched on had Korean natives lining both sides, bowing and chanting what sounded like manzai, or banzai. They were the first large group of people I had seen since my arrival and this was their way to welcome us. As we got closer to our objective, rifle fire could be heard ahead but the people still lined the road. The tempo of fire increased as we entered the city.

Seoul City

Sergeant McBride, at the head of our section, became a casualty by stopping a bullet in his shoulder and one in the leg. It was now our turn to experience war as blood began to flow. We moved further into the city and took shelter behind a row of houses, falling on straw mats that covered the ground, not realizing the mats were shrouds for Korean dead lying underneath. The dead resembled Korean natives we had recently passed on the road leading there. They were all clad in white garbs and were actually dumped in a sewer trench with mats placed over the top of them.

Enemy fire kept raking the ground near the edge of the house I was behind. I could hear bullets zapping the earth next to me. All I had to do was reach out with my hand or foot and the enemy marksman would guarantee a wound that would result in a Purple Heart plus evacuation to a safer area.

A figure bolted out of the house that we used for protection, to a partially fenced-in yard to my left. A shot rang out. Through missing slats in the fence, I could see a Korean native dressed in black. The Marine that fired his .45 at him, missed. The old Korean seemed confused. No one else fired at him so he scooted back to where he came from, probably thanking the Lord for sparing his life.

Dog Company was inching its way forward, using grenades to gain control of buildings offering resistance. In all this hell, I observed a very young Korean child playing in the road with a G. I. ration can. It's a wonder he wasn't killed with all the lead flying and hitting near him. A Marine scooped him up and carried him to safety.

Toward evening, I was told to take over Sergeant McBride's section. Dog Company was taking it's share of casualties. It got dark so we settled in for the night with firefights keeping everyone awake. In the morning, a wounded Marine was brought in by a group of Korean boys. I walked over to where he lay on the ground. A chaplain was on the scene praying over him, administering last rites. I asked, "What happened?" The wounded Marine was saying he got shot in the chest during the night, with a bullet wound that was close to being fatal. There were four boys involved, risking their lives to bring him in. Finally, as medics came to evacuate him, an emotional exchange took place. The wounded Marine wouldn't release his hold on one of the Korean boys. His gratitude was shown by offering to give the boy any personal possession he owned. Anything of value in his wallet now belonged to the boy, if he wanted to claim it. That Marine (Joseph A. Saluzzi) returned home and wrote a book called, Red Blood, Purple Hearts: The Marines in the Korean War. His night of terror can be read in his book.

Shortly after this incident occurred, information was received claiming some Marines were seen hanging in the city. Dog Company was assigned the mission to go out and recover their bodies. If I hadn't seen it, I wouldn't have believed it. There was a sailor who jumped ship and felt more at home in combat with Dog Company. He volunteered to accompany the unit on their recovery mission. They took off and upon return, were asked all sorts of questions about what they had encountered on this mission.

My section from Weapons Company, equipped with rocket launchers, was to stay close to the Dog Company Command Post. It was late in the afternoon when I was given a fire mission. I was told there was a large building that North Koreans were supposedly using for shelter. My rocket launchers were set up and fired some rockets into our target area. Dog Company Marines then went in to secure the building. Toward evening, a young Korean lad who spoke fluent English approached our defensive position, volunteering to help in any way if needed. He became a member of our unit throughout the night.

Wounded Marines became a major concern. How to get them out for life-saving treatment, became a top priority. There were no helicopters available. It was dark. Finally, some tanks arrived and a decision was made to use these vehicles to take them out. The wounded were helped aboard and I watched as they departed. I was given an area to defend for the night. There was a steady echo of gunfire taking place elsewhere, but our sector remained quiet.

Moving with Dog

In the morning, still attached to Dog Company, orders were given to move out. It was a beautiful autumn day. Past noon, we came to a streamlet and stopped to rest. Some Marines took off their shoes and decided to refresh their feet in the cool running stream. Morale skyrocketed with word that mail call would take place shortly. My section attached to Dog Company was told we would have to wait for our mail until our return to Weapons Company. I decided this would be an ideal time to walk over and talk to my old buddy Harry Griswald. He had just finished reading his mail and shared some of its contents with me. Harry and I received our transfer orders off the carrier on the same day.

Scuttlebutt began to circulate that Marines had acquired a new nickname. North Koreans called Marines "yellow legs that never sleep". This came as a result of probing operations by the enemy who found Marines awake and alert during their night movements. Our leggings were different from combat boots worn by the Army and clearly visible from a distance.

Dog Company moved out and we were back in the hills. Time passed. Finally, our column approached a village. The point Marines passed word back that we were coming into a minefield and to use extreme caution in where we stepped. Once past the minefield, we entered the village, coming to a schoolhouse with an adjacent clear field behind it. To its left was the start of a high ridge that we started to climb. Communications were by word of mouth, relayed from one man to another, down the line.

Mortar Barrage

We had gone a short distance up the ridge when I got word to move my section up to where the company Command Post (CP) was located. As I moved my unit up the hill, word came back up that some North Korean soldiers were seen in the village behind us. I reached the CP and was told to stay where I was. Suddenly, enemy mortar rounds began to hit behind us at the base of the ridge. Word came back up that the enemy had observers located in the village behind us. We took some casualties and a helicopter was called in to evacuate the wounded.

The chopper arrived trying to land in the schoolyard, but the enemy mortar fire shifted, with shells exploding near the chopper, preventing it from landing. While this was taking place, I decided this would be a good time to have a bite to eat. For some ungodly reason, I felt hungry and needed something to calm me down. I took out a snack from my field ration, and started to chew on a piece of food. That's when the mortar fire shifted again to the base of the hill, forcing those below to move up the slope hoping to escape the explosions causing damage in that area. The enemy shells were now hitting right on top of us.

A shell exploded close to me. I actually never heard it go off. As I lay on the ground, it felt like a sledgehammer bashed my head. I saw stars and blacked out. Regaining my senses, I took inventory to see if I was okay. The food in my hand disappeared. Enemy mortars were still raining down. I could hear shells hitting the ground around me. With some, there was a momentary pause when they landed, then a loud explosion followed. It was a terrifying experience. I heard cries of wounded around me. A lieutenant just above me had shrapnel pierce his helmet, inflicting a severe head wound. I now became aware of blood on my right knee. A corpsman checked it, tagged me, and told me I would be sent back to an aid station for further treatment. During this mortar barrage of September 28, 1950, my seagoing buddy Harry was fatally wounded. He went to assist a wounded Marine. As he was lifting him another shell exploded. Fragments of shrapnel ripped into his head and he died instantly. Whenever I hear the song, "Heart of my Heart", it brings back memories of Harry. It was his favorite tune. I can still picture Harry singing this song many times aboard the aircraft carrier. It's a shame his honeymoon had to end this way.

All the wounded were now assembled in an area near the road behind the ridge. While waiting for transportation to arrive, I was now exposed to the agony and pain that badly wounded Marines had to undergo. It was a traumatic shock for me, witnessing the physical harm the enemy shelling had inflicted on my fellow Marines. The lieutenant

26

with a severe head wound kept crying and moaning of terrible pain in his head. I was surprised at the number of Marines who kept crying and calling for their mamas or mothers. The worst was hearing rasping sounds made by perforated chest wounds, air hissing as they laid breathing and dying. The Chaplain who paid us a call earlier was nowhere in sight. During this time of pain and suffering, his services would have given peace and comfort to those in need.

It got dark. Finally trucks arrived to take the wounded out. We traveled through the city of Seoul to a hospital aid station. Corpsmen were waiting, checking wounded, and seeking weapons. I was surprised at their insistence on procuring pistols. They were like a treasured souvenir, if a casualty had one. There was a show of excitement if any were found. I had in my possession a Marine issue .45 pistol that was assigned to me. I was asked to hand it over to the corpsmen that were treating the wounded. I just wasn't about to part with my weapon without an argument. A discussion with the corpsmen convinced me that it was in my best interest to hand over my weapon, with assurance that it would be returned to me when I was released to return to my unit. I received treatment and was shown where to rest for the night. In the morning, I was told I was free to rejoin my company. I retrieved my personal gear, plus pistol, and now had to find my way back to Weapons Company.

I hitched a ride on a jeep heading toward 2nd Battalion 7th Marines. Our journey took us through the city of Seoul. It was daylight, which gave me a chance to view the terrible destruction war inflicted on the city. Streets we rode on had wires and rubble scattered everywhere, some buildings were totally destroyed, and others were just mere shells.

Road to Uijongbu

Along the way I kept asking where I could locate Weapons Company. I finally hit pay dirt when I was told they were on the road I was on, heading for Uijongbu. I finally caught up to them. They were marching on both sides of a dirt road with jeeps and trucks forming a convoy moving between them. I found my lieutenant. He told me to report to my section, now under Sergeant Lister's command. I made my way to where I belonged in the column, taking in the scenery as I walked.

The tranquility of our march was shattered as shells exploded nearby, with everyone hitting the deck. As I lay on the road, I could see a railroad embankment that rose above the ground a few yards ahead and to the right of this road. It looked like foxholes were dug on the forward slope, just waiting to be occupied. Some other Marines and I with the same idea got up and made a rush for the shelter these holes offered. I was lucky to reach one that wasn't occupied and crawled in. It was just large and deep enough to provide protection from everything except a direct hit. The volume of incoming fire increased as the convoy was the main target. Direct hits were taking a toll on vehicles and Marines. As the shelling continued, enemy shells were scoring hits on the embankment that protected me. With each explosion I could feel the earth tremble beneath me, and at times I felt dirt fall on top of me. Whoever dug these holes deserved a world of gratitude for the protection it offered from the deadly artillery shelling.

This was a scary experience. Just a day before, I had been exposed to deadly mortar fire. Now artillery shells were exploding around me. How I survived, only the Lord knows. The shelling stopped. We vacated our holes and returned to muster with the rest of the platoon. Damage was assessed. Corpsmen were tending to the wounded. There were two Marines from my platoon who sought shelter alongside the vehicles. Shrapnel from enemy shells exploding near the convoy resulted in serious leg wounds to both of them. They had portions of their limbs blown away.

My platoon leader survived a hair-raising ordeal, as shrapnel pierced his helmet without causing any injury. Without that tin pot, he would have been a goner. The helmet became a trophy he would take home as a souvenir. It would always bring back memories of the day he survived a North Korean artillery barrage with the helmet saving his life.

Now that the shelling ceased, everyone on the road started to move forward. Shortly we came upon one of our tanks; it had run over a land mine and was disabled. One of its tracks was blown apart. I checked the damage as we passed. No one inside the tank was injured. Daylight had turned to darkness and orders were given to dig in. Shelling by both sides kept everyone on edge throughout the night, the scariest time occurring when our rocket batteries opened fire. This was the first time I was exposed to rockets in action and I was impressed. Shells from a battery caisson unleashed a volume of fire I had never before seen. At first, I wasn't sure which side was responsible for the activity, until I saw the mass explosions going off in front of our perimeter.

The following day our unit was kept in reserve. It was during the night that I had another terrible experience. We were assigned to an area just to the rear of the forward units, with each foxhole on fifty-percent alert. I was paired with a Marine by the name of Marshall. We had agreed to a schedule on keeping awake. It was while I was dozing that I felt a blow to my head. Upon opening my eyes, I found myself starring into a barrel of a pistol. My partner had fallen asleep while a staff sergeant was checking positions. If that had been the enemy, I would have been history. The sergeant was trying to instill the importance of staying alert, even though we were behind the front line. The sergeant took a gamble with his move. I later talked to other Marines who were awake, telling me they would have blown him away except they knew who it was prowling. They weren't as trigger happy as were Marines on our second day ashore. I questioned my partner as to why he had placed us in such a dangerous position. He said he was so tired from lack of sleep that he must have dozed off. I told him I didn't appreciate the blow to my head because of his carelessness, not knowing if he also got the same treatment. We got away with one this time, hopefully, it wouldn't happen again.

At daybreak, we were on the road heading into Uijongbu. As we entered the city, I looked at the remains of what once had to be a beautiful city. The place was a total disaster. There was only one building with a partial wall standing. That was it. As we marched through the wreckage, I observed many dugout bunkers used by

Korean natives for shelter. Some had heavy timber buried beneath dirt to protect the structure from shellfire. Koreans standing near those dugouts watched as we passed by.

That evening, my section was detailed to man a roadblock on a road located a short distance west of an intersection heading north. Shortly after dark, we could hear a vehicle approaching from our rear. As it got closer, we could hear a verbal exchange taking place that sounded Korean. It hadn't reached the crossroads when a discussion took place among the senior NCO's, deciding whether or not to challenge the vehicle if it took the turn toward our roadblock. Finally, word was given that it was probably South Koreans who got lost and not to challenge them. Arriving at the crossroads, it made a turn and now was heading straight toward our roadblock. As it passed our position, I could see the outline of a truck disappearing into enemy country. The occupants of the truck never saw or knew we were there. I was located roughly twenty yards to the right off the road on a perimeter that tied in with heavy machine-guns on the right flank, extending and tying in with Dog Company, which occupied the road leading north.

It wasn't long before activity could be heard taking place in front of where we were located. It sounded like a truck was coming back. A discussion again took place, as to whether to challenge the truck as it returned. Once again, it was decided to exercise the same procedure as before to let the truck pass. As the truck passed our roadblock, we noticed that it had picked up some additional features. I could see a field piece hitched to it and more soldiers seated in the rear. On reaching the intersection, it turned north and went a short distance to where Dog Company was located. Shots rang out. A North Korean officer inside had a pistol that was now a treasured souvenir. Enemy survivors became prisoners. The truck became a motor pool possession. The field gun could easily have been responsible for some of the shelling we had received two days back near the railroad. This episode was embarrassing to my unit. Twice in a short span of time, we allowed North Korean soldiers to move freely past our roadblock without a challenge. This should never have occurred. During daylight, the identity of a truck and its occupants wouldn't have been a problem. At least by allowing them passage, they came back with the gun and more soldiers that were taken out of any further action against us.

In the morning we moved out, but instead of going forward, we changed direction and marched back through the town. This gave everyone a chance to again view the wreckage we had passed earlier. The weapons of war erased any existence of this community. We came to an area where trucks were waiting to take us back to Inchon. A rumor started to circulate that our job was finished and we were heading back to Japan.

At Inchon, we found shelter in an abandoned factory while waiting to be shipped out of Korea. Everyone had a chance to relax and catch up on needed sleep. Bartering with natives became a game. The most sought item was fresh eggs. We hadn't tasted any since our arrival and were still being fed field rations. Korean natives swapped eggs for cigarettes or candy. Time passed quickly. On 13 October, we were taken to the waterfront where our ship was waiting. It turned out to be an LST USS 973, manned by Japanese seamen. Rumors of going to Japan vanished.

Operation Yo Yo

As we left port on October 15, 1950, we received information that our 1st Marine Division would make a seaborne invasion on the beaches of Wonsan harbor, scheduled to take place on October 20th. Aboard ship, I shared a bunk area with a World War II veteran, who was one of the first Marines to occupy Japan. He met a Japanese girl, fell in love, and they ended up married. He was spending a great deal of his time lying in bed reading mail from home. This aroused the curiosity of some of the Marines sharing the same quarters. They asked him why he wasted so much of his leisure time reading the same letters over and over. That's all it took for him to unwind. He sat up on his bunk with letters still in hand and began to answer the question that was raised. It was a personal account of being married to a Japanese girl and their life together, going into detail about customs, respect, and love they both shared. We all just sat spellbound and listened while he narrated a beautiful love story. He was a sergeant with H & S 7th Marines. It would have been great if the story had a happy ending. As in every war, someone pays the ultimate price. The sergeant's dreams of returning home ended at the Chosin Reservoir. He died from enemy fire that riddled the vehicle he was in, as his unit was fighting its way out of the encircled town of Kotori.

As D-day got closer, a problem developed. Mines were spotted in the waters around Wonsan, resulting in a mine sweeping operation to ensure safe passage for the troop ships involved. This delayed our landing until the 26th of October. On D-day, we boarded Amtracs and started to move toward our landing zone. This would be my first amphibious landing against enemy defenses and I didn't know what to expect.

We hit the beach, but there was no opposition because Korean army units had raced from the 38th parallel and secured the area prior to our landing. Information that the landing zone was neutralized was withheld from us, until we came ashore. One Marine, a veteran of the war in the Pacific, leaned against the landing craft on shore and remarked, "Man, It sure is nice to light a cigarette and not have to worry about being shot or killed." Moving off the beach, I surveyed the enemy fortifications that were in place, waiting to repel any invading intruder. There were bunkers, trenches, and fire pits protected by barbed wire; also dirt covered enclosures to protect aircraft and vehicles.

Leaving the beach behind, we marched past a Marine Air group that was in place and operational. They got there before us. Good natured ribbing was dished out by the flyboys. No one took offense. It was great to be alive and the enemy was nowhere in sight.

We came to a road that took us into a town and stopped. While resting, I looked over the buildings that lined the road. It looked like they had escaped any major damage. Finally, word came down that we had a long way to go. Ten minutes breaks would take place every hour until we reached our destination. Twenty plus miles later, we arrived at a monastery, our bivouac area for the next few days. Lying around aboard ship with no physical activity had caught up with me. At the end of the march, lugging a full field pack, I ached all over and could hardly move. I imagine others felt the same. That night I slept on the floor and when morning came I was still in agony.

North to Chosin - November, 1950

After breakfast, my unit moved out on patrol to reconnoiter hills near the monastery. The longer I walked, the better I felt, and gradually all my pain disappeared. The terrain reminded me of home during hunting season, where I worked the fields to scare up pheasant or chase deer. We encountered no enemy. Our patrol ended, so we returned to the monastery and relaxation. It was a Catholic monastery with nuns in control. One of the rooms was used as a science laboratory. It was a shame to see the room a total mess with specimens splattered all over the floor and some in broken jars lying just outside the building. It had to be a diabolical person or persons to inflict this kind of destruction.

There was a variety of farm animals running loose. Sergeant Lister found a choice goat, decided it was time to change our diet, and butchered the goat. We used a bacon ration as lard in frying the meat. The share I received wasn't much, but it sure tasted delicious. It was only a matter of time before nuns became aware that some of their livestock had begun to disappear and notified officers in charge as to what was happening. The following day orders were issued. Anyone involved in killing livestock would be severely disciplined. I heard the remainder of our goat disappeared into a well.

To kill time, some Marines played poker. A Marine, by the name of Kotek, won a thousand dollars. He was the big winner in the game. There were reports of North Korean soldiers in the vicinity. Some had engaged Marine outposts in a shootout, so we were alerted to exercise caution around our area, especially at night.

The following day our regimental commander, Colonel Litzenberg, held a troop formation. He had some new important information he wanted to share with us. We were told about reports of Chinese soldiers crossing the Yalu border into North Korea. He went on to say this could very well be the beginning of World War Three and that trucks would arrive shortly to move us north to meet the new enemy. We waited until dark before the trucks finally arrived. We got aboard and headed north toward Hungnam. From there, we moved north to Sudong.

On November 1st, we reached our jump-off assembly area. That night lying in my sleeping bag, I felt something crawling over me. It could have been a snake or some other creature. It disappeared before I could make out what moved across my body. At dawn, we saddled up and marched north to relieve South Korean soldiers who supposedly were fighting Chinese. Rifle fire echoed from hills as we got closer. Rifle companies came under enemy fire as relief was taking place. South Korean soldiers who were relieved were observed coming down from the hills. Our section continued moving north, until ordered to climb a hill on the left side of the road.

We stopped at a Korean house and were told to form a perimeter around it. The house was protected by a stone wall and we used it for our defense. The wall was at least three feet high, encircled the house, and gave us protection without the need to dig foxholes. A battle raged nearby; heavy rifle fire was being exchanged. Enemy shells rained down, exploding close to the house, but none caused any physical or material damage.

That night, someone said we were surrounded. With all of the activity taking place nearby, hardly anyone slept during the night. In the morning, a Chinese soldier was seen running on a railroad located a few hundred yards to the right of the road below us. Everyone with a rifle opened fire trying to bring him down. The man fell down and just sat there. Marines from a hill above him ran down and took him prisoner. From our distance we couldn't tell whether he was wounded from our rifle volley or whether he stumbled on the railroad stones or ties. He may have been hit, which could have been the reason for him going down. He was the first Chinaman I saw. He must have carried a lucky charm to survive the volley of shots fired at him.

It was time to move. We saddled up, moving down to the road, heading north, crossing a small wooden bridge with a narrow stream flowing underneath, and coming to a mountain pass with cliffs on both sides. The road began to pitch upwards as we marched, approaching the stillness of death as we came upon the bodies of Chinese soldiers lying in the culverts on both sides of the road. They were probably on their way to engage us but were gunned down by our rifle companies occupying the high ground above. There just wasn't any place for them to seek shelter. They were the first dead Chinese I saw. There were so many of them laying head to foot in a row from one another; it looked like a domino effect scene where one man went down and the others followed. No one in my unit stopped to check the bodies to see if there were any wounded among them. There was always a chance some could have survived by playing possum. Further up the road, the terrain leveled out with reports reaching us that Chinese tanks were spotted dug in and waiting. We were the anti-tank platoon and this was our specialty. We therefore, expected to be given the assignment to take them out.

We proceeded forward, passing some of our artillery guns that were dug in shelling the enemy. A short distance from the guns, enemy shells started to drop and explode near us. I dropped to the ground for cover; the enemy guns were seeking to destroy the field pieces we just passed. Nearby a Marine from my section sought shelter behind a huge rock. A shell exploded in front of the rock. The explosion had a devastating effect on the Marine. He was now seen trying to dig a hole in the ground with his bare hands, blood dripping from his fingers as he cut himself on the rocks in the earth. This was his second shell shock experience. The first took place outside of Seoul. He was hospitalized at that time, but when rumors of the Division going to Japan surfaced, he voluntarily left the hospital to rejoin his company, never expecting to be exposed to combat again. The shelling stopped. We moved ahead to where the Chinese T-34 tanks were dug in. We got there too late; they were already destroyed by our air support. A halt was called; I could now sit down to relax alongside the road.

In the field to my left, I observed some Marines walking toward a stream and into it. A shell exploded in the water near them, I watched as some Marines fell. From where I rested, I could see other Marines rush to assist those in need. Word came to move out, so I couldn't tell whether those that fell were wounded or killed.

That night I found shelter in a dugout alongside the road. There was an older Marine with me, a reservist who was activated due to manpower needs. During the night, he kept telling me about this feeling he had that he wasn't going to survive the war. It was a premonition that wouldn't go away. I told him to forget about dying and to be more optimistic about surviving that he made it this far without a problem and his luck would hold. But all my effort to rally his morale was of no avail. He was adamant he was destined to die. He just didn't know how soon it would happen. Random enemy shells continued to explode in the area near us. Luckily, we lasted the night and we both survived. It appeared enemy troops encountered since the beginning of November had taken a terrible beating and were now withdrawing. Lately, resistance was light or none at all.

The next day, we boarded trucks and rode up a precipitous mountain road with no shoulder on the left side. It was scary just to look down the sheer drop to the bottom of the gorge. A careless move on the driver's part would carry everyone in the truck to certain death. Still a good distance from the top of the road, we had to cross a narrow bridge built against a sheer mountain wall, with only a footpath allowing passage on the side of the hill. The bridge was the only link connecting the road used to supply forward elements of the division. All vehicles, whether jeeps, tanks, or trucks carrying ammo supplies, field rations, you name it, had only this bridge to get across. There would be no way out for any vehicles if the bridge were blown away.

Schoolhouse Action

On November 10th, the Marine Corps birthday, we entered Kotori, There were no festivities to celebrate this special occasion. The weather turned bitter cold. We slept on the ground without digging foxholes. It seemed like the weather was getting colder each day the further north we moved. The night passed quietly. In the morning we advanced toward Hagaru. A plane crashed not far to the right of the road we were traveling. Some Marines went over to view the wreckage and to see if the pilot survived. Unfortunately, his luck had run out, adding another dog tag to the growing list of KIA's in Korea.

We advanced to a pass, setting up positions atop the ridge. It was dark and we bedded down for the night without digging in. We now had a new enemy to contend with; freezing cold temperatures came calling during the night. In the morning as we prepared to leave the ridge, some Marines complained of numbness in their feet and were told to rub them so they wouldn't freeze. That's when problems with frostbite surfaced. We still wore our original summer clothing that we had come ashore with, at Wonsan. We weren't equipped to cope with this new enemy. As I was coming down the hill, I saw one individual's hand. It had swollen to twice its normal size. This was my first exposure to a frostbite injury. As far as I knew, I had suffered no ill effects from the cold that night other then freezing my butt and feet. We had no winter clothing to protect ourselves from the elements encountered during the night. Unless winter clothing was issued immediately, there was no telling how long any one of us could survive in the frigid weather.

In leaving the area, little did we know that a major disaster would occur in this location in the near future. On the road again, we set out for Hagaru, crossing a frozen stream, arriving in town around noon. There were houses everywhere; a luxury we hadn't seen for a few days. Later in the afternoon, we were taken to a building where winter gear was issued. I got my shoepacs, parka, winter hat, vest, gloves, etc. The timing was a day late for some, but a blessing for the rest of us.

A plane had crash landed in a field not far from the buildings in town. Someone with a great deal of wisdom decided this would be an excellent location to build a landing strip and work commenced immediately. As luck would have it, we said goodbye to the houses and moved out to form a perimeter on the north part of town. We got there before dark and found the place loaded with standing haystacks everywhere. Since there hadn't been any sign of enemy activity for the past few days, we decided to use haystacks as shelter for the night. I crawled inside of one with my partner and spent the night here. Amazing how warm it kept us. The only worry was, no one dug any fox holes that night. It would have been hell without them, if the Chinese had come to pay us a visit.

A decision to man skeleton outposts the next day gave us a chance to move indoors. We occupied a schoolhouse a short distance to our rear. Relief took place every two hours. A heavy machine gunner named Miller came down with the flu and the runs. He was in bad shape, yet refused to be moved from his gun position to the shelter of the schoolhouse. Talk about a dedicated Marine; he had to be one.

During the night while I was in the house, a Marine ran in shouting that Chinese were seen entering the building. Someone said, "the garage." I was awake and fully dressed minus my parka when this took place. There was now a mad scramble taking place as everyone reacted to what they heard. I heard someone yell to vacate the building. It took only a second or two to slip into my parka, grab my rifle, and cartridge belt. I then heard the same voice yell to use the window to vacate the building. It took less than a minute and I was outside the building. Some Marines, on hearing word to clear the building didn't waste any time, vacating the schoolhouse, leaving shoes, winter clothing, and even personal weapons behind. I couldn't believe they would vacate the building dressed as they were in this frigid weather. As we prepared to take action against the enemy in the garage, my former foxhole buddy, Marshall, a physical fitness nut, vacated the house without his weapon or winter clothing. On being told Chinese were in the garage, he immediately dashed inside. Shots rang out; Marshall came out holding his arm. He was hit and blood stained his clothing.

This set the stage for everyone to fire into the garage. Some fired their carbines on full automatic just to saturate the building. A Marine officer arrived on the scene and had a couple of vehicles moved into position, using the headlights to illuminate the interior of the garage. Now the shocker took place. The garage was empty. Then how did Marshall get shot? I listened as officers conversed; saying the Air Force was alerted all the way to Japan, ready to supply air power if the need arose.

Sergeant Lister came over and told me to relieve a machine gun outpost. I got my gear together and set out to make a tag relief. The outpost was located on the extreme north part of town inside a haystack. Only a fire lane was clear. It was a cold, dark, brutal night with gusting winds adding a chill factor that made it a great deal colder. Shortly after my arrival, an explosion occurred near my gun position. With the excitement that took place at the schoolhouse just before coming there, and now the explosion going off near me, I didn't know what to expect, as I scanned the area around me trying to locate any movement. I finally decided it was a "bouncing Bettie" [grenade] that was responsible for causing the noise I heard. As I sat there freezing and daydreaming, I was jarred alert as another explosion took place. More explosions occurred near me. Each time one went off, I tried to see if any enemy were trying to infiltrate our perimeter. I couldn't see anyone and wondered what had caused the explosions. It could have been a small critter, or winds blowing hard enough to cause our booby traps to detonate, if that was what actually happened. I was stationed at what should have been a two man outpost, but there I was all alone and couldn't see where our next listening post was located. I wasn't even sure my machine gun would fire in the cold weather. It was a long, scary night. With daylight came my relief.

I now had to know what occurred at the schoolhouse during the night. There were two explanations. One was that whoever sounded the alarm may have seen a relief group returning to the building, mistaking them for Chinese. The second story I heard was that Marshall had been riding this tall rebel for quite some time. Seeing Marshall race inside the building, the rebel opened fire and those were the first shots we heard. If Marshall were killed with the Chinese in the garage, then there wouldn't have been any suspicion directed toward him. Unfortunately, absence of any dead bodies made the situation difficult to explain. I never saw the rebel after this incident. We had to admire Marshall's valor, willing to take on the enemy with his bare hands.

Our unit was relieved when another outfit moved in and took over our perimeter. This move was a real blessing. As we moved back to a house near a stream that was completely frozen, there was more activity taking place in this new location. One day some reporters came by, took some snapshots, and asked our names and where we lived. The photo I was in, appeared in my local newspaper back home. Sergeant Lister figured it was time to change our diet and cook up a special treat. All we had eaten since leaving Wonsan was a steady diet of field rations. Once word got out as to what was going to take place, everyone got involved. A large kettle was available in the hut. We pooled our assortment of field rations and in a dirt mound near the house we dug up some potatoes. All these goodies were dumped into our pot. There was a stove in the house. Someone started a fire. Our goulash was placed on the stove to cook; the hot meal in this cold weather was a treat with all credit going to Lister for his ingenuity. There was another benefit, too. The heat from the stove traveled into flues buried in the floor. It was like having our own electric blanket, with the floor radiating heat. That night I tried sleeping on the floor. It was nice and warm. At times it got uncomfortable, but it was better than being outside in the freezing cold.

Road to Yudam-ni

Our stay there was short. We had to relocate northeast onto a road leading to Yudam-ni and set up a roadblock. When we arrived at our designated area, I got my shovel out and tried to dig a foxhole. Instead, I found myself spinning my wheels. The ground was frozen solid. Each attempt to penetrate the dirt proved a failure. Finally, someone came over with long-handled axes. There were only a few to go around, so once a Marine with the axe broke through the frozen ground, it was then passed on to the next man in need. While we were digging, a bulldozer arrived and dug up a shallow pit just behind our foxholes, pushing dirt into a mound facing northwest. Once our foxholes were dug, it then became a challenge to see who could erect the best protection for the top of the holes. Finally, each hole had overhead cover. Some were better than others, depending on where the wood came from. It also helped in keeping some of the cold weather out.

It was Thanksgiving Day with rumors circulating we would be home for Christmas. A new offensive scheduled for the following day was to take us to the Yalu River. If the offensive was successful, a link up of the X Corps units with Eighth Army would hopefully end the war in Korea. That evening, we celebrated with turkey and all the trimmings. On the way back from the mess tent, I walked by our G-2 tent where a couple of old Koreans were being questioned. I watched as the officer in charge was told that many, many Chinese were in the hills, pointing in the direction of where they were seen. If that information was accurate, were they waiting to attack or was the information given to deceive us? I then left and proceeded to my home for the night. This would be our last day at Hagaru. In the morning we were scheduled to leave for Yudam-ni to join the rest of our battalion.

The night passed quietly. Early the next morning, a number of field ambulances passed through our roadblock heading for Yudam-ni. We were in the process of preparing to leave when some Marines were spotted running toward us on the road leading from Yudam-ni. As they entered our perimeter, they said Chinese soldiers were waiting in ambush and had shot up their ambulances. They were lucky to survive. I looked at the high ground in front of my foxhole. Chinese soldiers could be seen moving south in the direction of the road where the survivors had just come from. The Koreans the night before were right on the money in pin pointing where the enemy was located.

Fox Company from our battalion was located a few miles up the road. Information of Chinese cutting the road between Fox Company and us was immediately reported to our C.O. Shortly, a .75MM recoilless rifle was brought up. They positioned their weapon and got ready to shell the moving enemy. We, in turn, were alerted to move away from behind their gun because it had a tremendous back flash that could kill or severely injure anyone close to its blast. While waiting for word to move out, we watched the gun crew shell the moving Chinese.

Our orders were to relocate to Yudam-ni. With everyone saddled up, we marched to a truck assembly area and were told to climb aboard. It didn't take long for the trucks to move out. Everyone in our section was concerned what would occur once we got near where the Chinese ambush took place. All the trucks we rode were covered with canvas tarps. Passing our departed roadblock, tension was high. We hadn't gone far when the trucks stopped and we were ordered to dismount. If the ambulances hadn't tested the road earlier that morning, it would have been our trucks that now would be exposed to hidden Chinese fire. God only knows how many casualties were prevented by the earlier encounter.

The word came down that we would make an assault to destroy Chinese positions blocking our way. Lieutenant Henderson, .81MM mortar platoon W-2-7, led the attack. As we moved out in the assault, Chinese troops were ready. Lieutenant Henderson had taken only a few steps when he fell from a bullet wound to his stomach. There were other casualties and the attack stalled. There was a great deal of confusion. New orders were issued, and Sergeant Lister was told to have his section occupy the north side of the road. As we moved off the road, a Marine, who was a heavy machine gunner, was being assisted to the shelter on the hill. He had suffered a severe head wound. His bandage was stained with blood and he kept moaning with pain.

Sergeant Kraus, a section leader in our platoon, was ordered to clear a hill on the right flank of the road. From our position, we watched and waited as Sergeant Kraus led his section up the ridge. There was no preparatory mortar or artillery fire to aid the unit as it advanced toward the Chinese. Near the crest of the hill, the unit ran into a buzz saw. Sergeant Kraus was killed instantly and the rest of his section was pinned down taking casualties. Marines in the attack were having problems with their weapons firing. For example, one Marine, I believe his name was Corporal White, ran into a Chinese soldier who had a Thompson submachine gun. The Marine's carbine jammed and would not fire. The Chinese soldier's gun also failed to fire. Another Marine named Miller coming from behind, also with a carbine, had the same problem. His weapon jammed. They bayoneted the enemy soldier, claiming the Thompson as a prized possession. Later one or the other could be seen carrying the Thompson. It was evident that the unit that had been led by Sergeant Kraus was in trouble and needed help.

Sergeant Lister was ordered to take his section and help extricate the wounded and the rest of Kraus's section off that hill. We went over and found the unit moving down, and aided the wounded that needed help. Kotek, a big winner in poker the month before, was shot in the groin area. Hopefully for his sake, he was able to retain his poker winnings. Once everyone was back on the road, a decision was made to forgo any further action against the Chinese. We returned to Hagaru leaving all our dead behind. The roadblock was again our home for the night.

The following morning, all Marines in Weapons Company were assembled and told we were going back to dislodge Chinese soldiers blocking our passage to Toktong Pass where Fox Company was dug in. This time there were no trucks to ride. We had to dog it on foot, so it took a while before we got into position for the new attack.

We approached the Chinese positions with a tank in support on the road. Our orders were to take the ridge that Sergeant Kraus had been killed on. Everyone was told to fix bayonets. We deployed in attack formation and started up the hill. It was a textbook assault as .81MM. mortars walked us up the slope. The Chinese were waiting for our attack, and as we got closer enemy riflemen opened fire. Bullets were flying, seeking victims. The tank on the road was dishing out its share of firepower.

As we got closer to the top, I heard Ballard Lawing to my left yell, "I'm hit. I'm hit." As he fell to the ground, I dropped alongside of him. He kept yelling, "It hurts. It hurts." I asked where he was hit and he said in the back. I checked the back of his parka and saw what looked like a bullet hole but no blood. I told him I would take a look to see how bad it was. I then proceeded to peel the layers of clothing off his back. As the last piece of underwear was removed I relaxed upon seeing the extent of his injury. I now understood why he was screaming when he got hit. I told Ballard (we called him Tennessee Toddy) that this was his lucky day. The bullet burned the flesh across his back, leaving a red streak similar to a whiplash. He felt the pain as the bullet burned his flesh. If that shot were fired a second or two sooner, Ballard would have been in deep trouble. I was the third man on the extreme left flank going up the hill with Ballard and Douglas was to the left of me. The bullet that hit Toddy had to come from the area of the road where the tank was located. Chinese to the left of the road were receiving fire from the tank and had a clear view of our move up the hill. They were trying to take us out. Now that I knew Toddy was okay, I turned my attention to where all the shooting was coming from with bullets still zipping by.

I had cleaned my rifle the night before with gasoline. I wanted to be sure all the oil was removed from my rifle to prevent any malfunction. I

44

didn't want to find myself face to face with a Chinaman and having a jammed rifle like Marines had experienced yesterday. We were now in a position where Chinese grenades were raining down on us. I watched one Chinese soldier get up and toss his grenade. I zeroed in on him, but he dropped before I had a chance to fire. I waited, thinking he would try again. Sure enough, he didn't disappoint me. This time I was ready as he stood up, brought his arm back, and tossed his grenade. I fired a few shots, watched him fall, fired a few more shots at him, and then my weapon went dead. My God! What happened there? Now I was in the same predicament as Marines the day before. Would I survive the next few minutes? I checked my carbine, trying to find what had caused this to happen. I was astonished to find the operating rod had come off the bolt. I never heard of any such incident ever happening before. I didn't panic, but rather, worked quickly to reassemble the rod unto the bolt. With the carbine assembled, I now pressed up the hill, hoping my weapon wouldn't fail me again. Above all the noise, I could hear the commanding voice of Sergeant Lister, a World War II veteran, yelling to those around him to look out for foxholes. "If you see any, use grenades on them", he said. I could hear fellow Marines yelling, "I got one. I got another one." It sounded like a turkey shoot. As I got closer to the top, I saw a Chinese soldier a few feet from me. I quickly got a shot off at him. The rifle aimed at his head didn't fail me. I walked over and looked at the man I had just shot. With all the lead flying the scene could have been reversed, but it wasn't my time.

"Keep moving, keep moving" were the words that took me away from the fallen enemy, as we proceeded to take the crest of the ridge with hardly any loss. Once atop the hill, an eerie silence took over as the shooting stopped. There was a great deal of excitement displayed by Marines following this successful assault against the enemy. This was a new day and the hill that posed a problem the day before was now in our possession. I looked over the bodies of dead Chinese and noted clothing they wore was different from the ones we encountered earlier. These wore what looked like a faded white to blend in with the snow. The ones we had encountered earlier wore a standard brown color. They also wore sneakers in this cold weather. I don't know what the rationale was behind that; there was another surprise. The grenades they had thrown at us had caused very little damage. They were wooden concussion grenades, with wood splinters causing some of the wounds.

I went over to where Sergeant Kraus went down the day before. A bullet hole could be seen on his forehead dead center above his eyes. As he lay there, it brought back memories. I was one of three Marines sent to secure berthing for our company aboard the USS Bayfield at San Diego Harbor. We arrived a day before the rest of the unit was to arrive. There was liberty that night, but only two could go. We drew straws to see who would be left behind. It turned out I drew the losing straw. A short time later, Sergeant Kraus came over and asked if I really wanted to go. I told him I had never seen this part of California and since this was our last night there I really wanted to go and see what the town had to offer. That's when he told me I could go in his place and he would stay behind. I'll always remember him for his generosity that day.

Back to Hagaru

With the ridge in our possession, I expected to hear orders to sweep the flank toward the road. I watched as the officers involved in this operation held a conference to decide our next course of action. Shortly word came to relinquish the ground we just took. They made a decision to return to Hagaru. A golden opportunity was at hand. Chinese surrounded Fox Company, located four miles from Hagaru. We had punched a hole in the ring and now failed to take advantage of our success. Maybe our need was greater at Hagaru or else we may have run short on ammo. I'll never understand why we didn't press further. We had taken some casualties in this assault and one Marine that wouldn't be going home was the Marine with a premonition. He paid the supreme price that day. I can still recall the night we spent together. He just knew it was going to happen and he wasn't disappointed.

On returning from the ridge, there was a gunny sergeant waiting for us on the road. I overheard him saying to another Marine Non-Commissioned Officer that he just didn't have it anymore. That's why he didn't participate in the attack with other members of his platoon.

Once inside our perimeter, we found reporters waiting to hear all about our encounter with the Chinese, trying to extract as much information as they could. Maybe our episode made the news wires back home that night or else they needed information for Marine records.

Marine engineers worked day and night building a landing strip with floodlights to assist them at night. The Chinese attacked Hagaru from the southwest and northwest of our position. It started before midnight. When the Chinese attack began, I could see muzzle flashes in the dark with bullets whizzing overhead or slamming into the bank made by the bulldozer. Shells were dropping and exploding near our foxholes, but none causing any casualties. We expected them to hit us from the east, but coming in on the road leading to Fox Company, they hit Fox instead.

A Marine air group had erected tents across the road to our rear. Some Marines asleep on cots were killed in their sleeping bags when the Chinese fired the first shots. It happened so suddenly, it caught everyone by surprise. As the battle raged, Sergeant Lister ordered a member of my unit to fire a 3.5 rocket at a house, which appeared to have some movement around it. The rocket fell short. I was then given a chance to see if I could do better. Call it luck or whatever, but I scored a direct hit. The house caught fire and lit up the area around it.

We had no idea how bad the situation was elsewhere in or around Hagaru. During the attack, a forty yard gap was breached in our defensive line to our left. Service personnel in the rear were quickly assembled and rushed to plug the opening. They not only succeeded in disposing of the enemy soldiers who penetrated into the perimeter, they also closed the hole. The battle lasted all night with reports that at least twelve hundred bodies were stacked up in front of the defensive positions by morning. The Chinese withdrew at daylight to regroup. The number of Chinese casualties must have been tremendous. The rate of wounded almost always exceeded the number of dead left behind.

In the morning I took a walk across the road toward the Marine Air Wing tents. From there, I looked toward the area leading to Yudam-ni and was surprised to see a fairly deep trench running out into the open fields. I didn't see anyone covering that area. I don't know if anyone in my section knew it was there. The Chinese could have poured through that trench had they attacked from the east, but they chose different sectors of Hagaru to attack instead.

A Marine supply column coming from Kotori to Hagaru was ambushed at the pass where we had our first experience with frostbite. That area became known as Hell Fire Valley. The column was named Task Force Drysdale, after the British Marine Lieutenant Colonel whose mission it was to reinforce the garrison at Hagaru. It ran into trouble when Chinese ambushed the convoy, resulting in hand-to-hand combat all day and night with only a small group making it to Hagaru. The ones making it were from George Company of the First Marine Regiment, including Lieutenant Colonel Drysdale, and a number of British Royal Marines. All the others were killed, taken prisoner, or managed to find their way back to Kotori.

Engineers worked around the clock on the airstrip, never stopping until it became operational. Now all the units at Chosin were surrounded. A Marine in our company activated as a reservist, received orders to return home. Having packed all his gear, he waited for transportation out of Hagaru, only to find the road closed. He was trapped with the rest of us. I never did find out if he made it back safely to his family back home.

Yudam-ni Marines

Fox Company, occupying a key ridge location, was cut off and heavily attacked from all directions. Marine units at Yudam-ni engaged Chinese in all key positions. Yet for some reason, the Chinese spared our sector of Hagaru from a full assault. Maybe our encounter, with the enemy the past couple days, played a key part in our area being by passed with a massive attack.

We were told those Marines at Yudam-ni made up of the Fifth and most of our Seventh Regiment were under constant attack by the Chinese and were now ordered to fight their way back to Hagaru. They had to disengage from the enemy attacks against them, turn around, and clear the Chinese blocking the road to Toktong Pass where Fox Company held a vital piece of ground. Once this was accomplished, there was one more obstacle to overcome, to recapture the ridge area where we had our successful assault. The Chinese had reclaimed the ground after our withdrawal.

Reports came in that a link up was made with Fox Company and all the units were now headed our way. During the time Fox Company was surrounded, they were re-supplied with airdrops. Barbed wire fencing was an item dropped to help the defenders. Wire was strung and used to hinder the Chinese attack on foxholes in defensive positions. Hagaru was also surrounded, receiving airdrops, until the landing strip was ready for use. Royal Marines who had fought their way through the ambush at Hell Fire Valley were chosen to clear Chinese from terrain between Hagaru and Fox Company. They had no problem doing it.

Since we were dug in on the road leading to Yudam-ni, our unit was the first to see a Marine column approaching our road block. Marines entering our perimeter looked tired and worn, yet they marched with pride and determination after having fought their way through roadblocks manned by hundreds of Chinese soldiers. In the walking column were many limping, wounded, or frostbite cases. Others were wrapped with battle dressings to protect injuries inflicted by the enemy. Many walking Marines were hanging onto the sides of the slow moving vehicles for support. They needed that extra tug to keep them moving. Dead Marines were tied down to hoods, fenders, tops of vehicles or anywhere a place could be found. The dead weren't going to be left behind for the Chinese to ravage. I spotted Marines from my company who were in the mortar platoon and had survived the ordeal. They were just happy to make it to Hagaru.

Morale took an upward jump, now that the Fifth and Seventh Regiments added to the defense of Hagaru. The engineers, who worked day and night, did a magnificent job by making the airstrip useable for the wounded that were now at a place where air evacuation was available. Planes were landing and taking off with as many casualties as possible.

From the Yudam-ni Marines, came stories of hardship and glory as they fought through the Chinese hordes, which were determined to annihilate all of them. With temperatures ranging twenty to thirty degrees below zero or more, frostbite played a major role, affecting all the troops at the Chosin Reservoir. Almost everyone exposed to this frigid weather suffered some degree of frostbite. The more serious cases were treated and tagged to be flown out of Hagaru.

That night we had to relocate to a new position, located due west a few hundred yards. A South Korean outfit tied in on our right flank. Actually, they were positioned forty to fifty yards forward of our defensive line. As soon as it got dark, the Koreans opened fire and continued to do so all night long. No one slept. Finally, the Koreans ran out of ammo, coming to us asking for a new supply. We had none to spare so they went elsewhere. They had no fire discipline and their action kept everyone on edge.

Due to heavy losses suffered at Yudam-ni and Fox Hill, new replacements were flown in to replace the dead and wounded. Sergeant McBride, my old section leader who was wounded at Seoul, was one of them. I spoke to him the first day back. He didn't appreciate coming back to our predicament. I saw him again that night and then he just disappeared from sight. Marines weren't the only ones trapped at Hagaru. Some Army personnel that had made their way into Hagaru were also involved in the same predicament. Their numbers helped our manpower situation.

A mess tent was setup, giving some a chance to savor a hot cooked food. The only visit I recall making to the mess tent was to sample a breakfast menu. I was given a small cereal container, had milk poured into it, and then walked out of the tent to eat my meal. That was a mistake. Almost immediately the milk in my cereal started to frost up. It had to be cold for that to happen so quickly. Health conditions weren't any better. My nose ran all the time, causing my moustache to have a continuous frosting caused by the frigid cold. I couldn't remember the last time that any of us had taken a bath. Living conditions were terrible, especially since we spent all our time outdoors in freezing weather.

It was time to move again. We were given a roadblock assignment located southeast on a road, adjacent to a railroad intersection. Gasoline drums were positioned on the road to our front. A tank supporting us had orders to set the drums on fire in case of an attack at night. This would supply the visibility needed. On the other side of the road began the rise of a large hill occupied by Chinese troops. The railroad embankment provided our protection and shielded a tent with an oil burner inside. Marines on watch at night came in to warm up, huddling close to the oil burner to absorb the heat. That didn't help. It was so cold, the stove couldn't emit enough heat to keep the tent warm. The tent did, however, offer shelter from the cold gusts of wind blowing outside.

The first night at our new location Sergeant Lister came around asking for volunteers to scout the area in front of our perimeter. The area was covered with a heavy blanket of snow with the temperature way below zero degrees. In this weather everyone waited to see who would be the first to step forward. There were no takers. After freezing all day, we were now being asked to suffer more exposure to the bitter cold and it was more than some could endure. Had he just come out and selected whom he wanted, that would have solved the problem. Seeing no takers, rather than force anyone to go out and weather the cold out in the open, Lister decided to cancel the patrol. The freezing night played a major role in his decision. There were enough frostbite cases without causing more.

Army Stragglers

Information was received to anticipate Army stragglers coming through our lines. During the night, some did make it. I was selected to escort the first survivors to our first aid station. On arrival at the building used for medical treatment, the Regimental Doctor handed me a flashlight. He had a job assignment ready for me. I provided the light while the medical officer checked out the wounds. Every soldier checked had frostbite. Leather shoes were frozen solid to their feet. One Army soldier had a bullet hole in his shoe. The foot was frozen, so it wasn't bleeding. The medical officer had surgical scissors in his hand ready to cut, but instead tagged the man for air vac. There was no way he could aid the man until the foot thawed out. A Korean soldier with this group had a good portion of his upper leg blown away. How he survived without bleeding to death, I'll never know. After all the patients were treated, I returned to my unit.

The night passed and more Army stragglers were coming in that morning. One Army survivor, seeing we had a Chinese prisoner, grabbed a rifle from a Marine and shot the unarmed captive, killing him. The execution by the triggerman was revenge for the carnage and hardship the Chinese inflicted on him and members of his unit. He was one of the lucky Army soldiers to reach our perimeter. Most that entered our position were without weapons. I don't believe the survivor was aware of what he had done by his action. Prisoners as a rule were important for information purposes. Whether this slaying deprived the intelligent section of worthwhile information will never be known.

After this incident, Intelligence informed us that some North Korean civilians had agreed to assist Marines in gathering information about activities taking place in front of our positions. Their identity was to be a rolled up newspaper. It wasn't long after this information was received that some Koreans approached our roadblock, taking the rolled up newspaper out of their coat pocket to let us know who they were. This system brought back information that some American soldiers were badly wounded in a house a couple miles up the road and needed help.

Koreans with sleds resembling those pulled by Alaskan dog teams went out to bring them in. The ones brought back were all napalm victims caused by American planes. They were completely covered in blankets or quilts except for their face area. I looked at the horrible burns that were visible and could almost feel their pain. They had survived, but for how long?

Breakout to Kotori

East Hill

Our position was on the low ground. The Chinese held the high ground. They had taken the hill away from the Army unit that was assigned to hold it. This ridge was known as East Hill. A vehicle with a .50 caliber machine gun mounted on its rear was brought up to the road behind the tank. Chinese could be seen standing around on the ridge. Once in position, the gun opened fire; the Chinese disappeared beyond the crest.

Plans were to evacuate Hagaru and move south to Kotori. There was an ammo dump behind us containing hundreds of 3.5 rocket shells. A decision was made to use up all the ammo before our breakout to Kotori. Our rocket launchers were brought up and it now became our job to use up all the ammo before leaving. We proceeded to fire at the hill, blasting the area where the Chinese were seen standing. The launchers were also arched to act like mortars, where shells would reach a high trajectory and then come down like a mortar round. Those we fired over the top of the hill.

Just as everyone was enjoying shooting and using up the ammo supply, orders came to cease fire. Someone decided there were just too many shells to get rid off and not enough time to do it. Thereafter, the dump was rigged with explosive charges by engineers and blown up. We'll never know how effective our shelling was because we couldn't see beyond the ridge. Maybe some shells did score a kill or two. It was our job to waste as much of the ammo time allowed and in the process everyone that wanted got a chance to spend time firing the rocket launcher. The breakout to Kotori was to take place at 4 a.m. the following morning. The Chinese had cut off Hagaru from Kotori and had troops in position all along the eleven miles in between.

That night, our unit was moved to a new location near a food supply dump, giving everyone a chance to scout the goodies and take what they wanted. The food supplies were to be blown to prevent the Chinese from getting at them. There was no way all these supplies would be destroyed. There wasn't much that could be taken, since all the food in cans containing liquids were frozen and worthless. The most acceptable food was candy. I loaded my pockets with Tootsie Rolls, figuring this would last until we got to Kotori. No one slept that night.

We were assigned to Easy Company with Lieutenant Bey in command. Dog and Easy Companies were badly decimated at Yudam-ni. Our addition to Easy Company helped beef up the manpower void. Sergeant Lister, for some reason, changed his shoepacs for his old boondockers. Seeing all the Army frostbite cases wearing the same shoe, I thought he made a bad decision to switch. I guess he had his reasons for making the change.

Word went around that maybe the Chinese would allow us to leave without a struggle. I was hoping the rumor was right. The day before, there was talk that the atomic bomb would be used to break us free. No one actually knew how serious the situation was. There never was any talk of surrender.

Early on the morning of the breakout, we marched toward the head of the column. The lead elements with vehicles on the road were catching hell from Chinese on East Hill. A section of .81 mm mortars was put into action but needed shells that were still packed in their cylindrical containers. Some Marines in my section, including me, pitched in helping to break open the tubes, so the guns could have shells to fire on the hill where the Chinese were dug in.

The enemy fire took its toll. Webb, a reservist from Pennsylvania who was a couple of yards from me helping with the mortar ammo, died from a bullet to his back. A heavy machine gunner had a slug pierce his cheek and exit through his mouth. He also stopped a slug in the leg. With all the Chinese fire coming at us trying to put us out of action, we proceeded forward, passing vehicles on the road and descending into a draw, facing the hill that was causing all of our casualties. I saw a heavy machine gunner holding a barrel taken from his gun. It had gotten overheated from constant firing and the front portion was now bent out of shape and just a piece of junk.

A tank was positioned in the draw near the road with all its guns blazing at the hill. Lister walked over to the tank, picked up a phone, and talked to the crew inside. That was the last time I saw Lister. I moved further south in the draw close to the road away from the tank. That's when air support arrived. B26 bombers were being directed toward the hill. I watched the planes coming in so low that I could clearly see the pilot and gunner in his turret firing at the ridge. Rockets followed and finally napalm dropped and exploded.

We got by this obstacle and had gone a short distance, coming to a railroad trestle. Underneath, a Chinese soldier was lying down on a bed built off the ground and covered with blankets. Shots were fired at him and then it looked like vapor or steam was rising from his body. I wonder what caused that to happen.

We got back on the road and were making some headway when again, we came under intense enemy fire. Bullets whizzing by, Marines were being hit in front and to the rear of me in my column. The casualty toll was heavy. I dropped behind a small mound and could hear bullets zapping the ground in front of me, but I couldn't see the enemy. There was a house a few yards away on the other side of the road that Lawing and I could see. We decided to try and take it out, but then saw other Marines already there leaning against the side of the house. Before we could make a move to join the Marines near the house, word came to move toward a riverbed on the right side of the road. As I moved toward the riverbed, I checked some of the men who were hit. I couldn't tell if they were dead or alive. They had turned a pale, whitish color. We had to leave them behind for the wagon train coming behind us. They were near the road and would be picked up and treated by them.

As I was walking in the riverbed, a shell exploded near me and I felt a tug at my parka. I looked down and saw a portion of my sleeve had disappeared. We moved further south in the riverbed. I was saddled down with full gear, yet as I walked, I experienced a new surge of energy. I don't know where it came from. There were so many events happening around me that I didn't have time to be scared, yet I couldn't explain this feeling that came over me with bullets flying and shells exploding near me. It seemed like the equipment I carried felt weightless and I found myself walking like I was in a daze. In the riverbed, we were protected from small arms fire by the high bank. We made many stops, while we waited for rifle companies to clear the Chinese off the ridges on our flanks. Some lasted hours. The cold weather made the stops a misery; at least walking kept the adrenaline going.

Hell Fire Valley Again

It got dark and we moved back onto the road. We came to the area called Hell Fire Valley. It was the place where we had had our first frostbite encounter. We had already heard what had taken place in this area before from survivors who made it to Hagaru. I could see dead frozen bodies everywhere. The area itself looked like a junkyard. There were trucks, jeeps and tanks mangled and destroyed, blocking the road and off the road. What had once been a part of a Marine convoy loaded with supplies and manpower mostly lay in ruins. Would we pay the same price as the dead that lay before us?

I could hear Lieutenant Bey yelling to keep moving. I couldn't believe it. There didn't seem to be anyone around. Where were the Chinese that had destroyed this convoy and the men with it? The temperature was frigid. Maybe the enemy troops were asleep or trying to keep warm in some bunkers, deciding to wait until morning before engaging in any battle with Marines. They may have miscalculated that Marines wouldn't attack at night and chose to rest instead.

The Chinese weren't the only ones that used darkness to gain an advantage over the enemy. This allowed us to slip through the mess blocking the road. Once past the carnage, we got back on the road heading south. It was still dark when we heard a friendly challenge by Marines at a roadblock. What was left of us had finally made it safely to Kotori. It had taken us nearly a whole day to cover the eleven miles from Hagaru and we were lucky to survive. We moved into the center of the perimeter where tents were erected. Someone showed us which tents to enter. I hadn't slept in close to two days. I felt tired and dozed off. I couldn't have slept long when I felt someone shaking me, saying I had to get up and muster outside.

It was still morning and what was left of our unit was told we had to go back to Hagaru. The men and wagon train following us were drawing heavy fire. All available manpower was needed to help rescue the vehicles and personnel coming out. We started back, covered some distance, approaching high terrain on our right flank. We could see men and vehicles coming toward us. We moved closer to the column. Rifle fire could be heard in the immediate area. Off the road to our left were British Royal Marines coming toward us. I watched as they moved with caution and discipline, dropping to the ground seeking shelter, offering less of a target to the enemy. In contrast, U.S. Marines were more daring, walking upright, keeping a distance between each other, ignoring any enemy fire. We could now see troops and wagon train coming out in force. The bottleneck had been broken. Word came to turn around and return to Kotori.

Later that night, we were informed that our unit would again spearhead the attack out of Kotori in the morning. Once this bottleneck was broken, we could then proceed to the safety of Hungnam. The bad news was that the only bridge out of Kotori had been destroyed and the Chinese occupied the high ground around it. Replacement sections to rebuild the bridge were to be parachuted into Kotori and installed by engineers, once the enemy was removed from the high ground overlooking the bridge area. It had taken more than a day to reach Kotori, with many delays and no sleep during the entire breakout. It was a comfort to be assigned to a tent for the night, knowing security was taken care of while the troops got some needed sleep for the attack that would lead us to safety.

Human Chain

Early the following morning, we assembled on the road leading south and moved out, climbing the hill on the right flank of the road. Almost immediately shells came down, exploding to our right and behind us. The ground was covered with snow, making it difficult to climb. We moved forward. It was bitter cold and snow started to fall.

After a while, visibility was hampered by the heavy snowfall. This made the climb that much slower. Finally we halted, waited for what seemed like hours, and then started to move again. The snow coming down made it impossible to see. Word came down to hold hands, so as not to lose anyone. We now became a human chain, moving at a snail's pace with only the lead man knowing where we were headed. I had my doubts about that, especially since none of us could see anything but the snow that was engulfing us. It was so bad that I couldn't tell whether it was day or night. Finally word came down to stay together and bed down. We couldn't go anywhere in this blinding blizzard. We crawled into our sleeping bags and waited.

The Chinese had to be close. We couldn't see them because of the heavy snow that was coming down, but they kept up a verbal communication system that could be heard as they talked to each other all night long. Sergeant Lister, a former China Marine, might have been able to understand what they were saying, but he was no longer with us. A machine gun broke the silence and small arms fire followed. We couldn't see anything, nor did we know what was happening. It was so cold that everyone worried whether our weapons would function. That was the most serious concern affecting us. Would our rifles protect or fail us in this frigid weather? We decided amongst ourselves to use grenades and knives, if the voices we heard came near us. No one slept during the night.

Blessed Snow Cover

In the morning, the weather cleared and the snow stopped. We were completely covered by a white blanket. On the next ridge near us we could see Chinese soldiers still jabbering and shaking the snow off their blankets. With this chore completed, we watched them move off the hill. Even though they held the high ground, they never knew how close we were. The snow cover prevented our disclosure. Whether our weapons would have actually functioned in this frigid weather was never tested. There was no attempt made to fire at the Chinese on the hill.

With their departure, we got out of our sleeping bags and had a chance to look over the hill we occupied. There were a couple of bunkers on the slope that we checked out and then waited for word to move out. It was close to noon and we still hadn't moved. Tootsie Rolls that I salvaged at Hagaru were the only food I had. Water in our canteens was frozen and useless, so we built a fire and heated snow to make coffee. This worked fine, but about this time, someone on our ridge became aware of strange movement below the crest of the hill the Chinese vacated earlier.

When the Chinese left, another Marine unit had moved in and occupied the ridge. A radio message was sent to the Marines on the other hill to have them check out that particular area in question. Shortly a rifle squad moved down to check out the problem. As the unit moved down the hill, I could see one, two, three Chinese soldiers coming out of a bunker with their hands up. In a flash, the Marine leading his squad could be seen falling. Shots came out of the bunker, hitting the ground we occupied. I could hear bullets hitting the frozen ground around me. We were down in a valley on the forward slope with no place to seek cover. The Marines on the other hill reacted. The Chinese, who elected to fight, died quickly, as grenades and rifle fire accounted for ten soldiers who occupied the bunker. The ones willing to surrender also perished.

Inside was found a large assortment of weapons, plus a clear view of the road below. I can't imagine the number of lives and vehicles that were saved by taking out this bunker. I also consider myself lucky--the whole time we were on the forward slope in the open; they never fired at us until the end. They had such easy pickings and never took advantage of it. With the firepower they had in the bunker, the Marines counted a total of twenty rifles and automatic weapons inside. They could have destroyed most of us, if not all. How many times during the early morning hours was I along with the others, a target in the sights of the bunker occupants? If it hadn't been for all the snow that had fallen the day before, making visibility impossible, I'm certain that my unit would have walked right into the Chinese encampment. I shudder to think what might have taken place. The blizzard was a blanket that protected every one of us from a killing confrontation. We may have suffered physically from the freezing cold all day and night, but the snow was a blessing and we all survived.

Bridge Repair

Shortly after this incident, word came to move out. We came back down to the road where I was surprised to see a large number of Chinese prisoners, which had surrendered during the night. As we moved out, we had to avoid the Chinese dead lying in the road that were run over by tanks. They looked like crushed pancakes. In front of us were hundreds of North Korean natives moving in the same direction, fleeing to escape the Chinese. We had to clear them off the road to enable our wagon train to move with us. We got to a certain area and stopped. Word came down that Marines from the First Regiment attacking toward us, had cleared the high ground near the bridge area. Now we had to wait until the engineers put the bridge spans in place. Hours went by. Nighttime came and we waited. Finally, trucks arrived and we climbed aboard. We came to the bridge and the trucker took it nice and slow. It was scary with shells exploding and rifle fire still taking place nearby. The truck started to speed up after we were safely across the bridge. We were on our way to Hungnam. I must have fallen asleep.

Tagged for Evacuation

I remember coming to a first aid tent and leaving my pack and rifle outside. The corpsman looked at my feet, said I had frostbite, and told me to massage my feet. I followed his instructions. After spending some time doing this, I put on some clean socks and left the tent. I went to pick up my pack and rifle but someone beat me to it. I had no other alternative, so I helped myself to someone else's pack and rifle. I had a guilty feeling doing this, but then someone else took mine with all my personal articles that were dear to me. I was keeping a daily log of events I was involved in and now it was gone, but I wasn't leaving without a pack or rifle.

I walked over to where Marines from my battalion were assembled. Shortly, roll call was held to count the number of Marines who made it back safely. I heard 25 were left in Weapons Company and 79 in the whole battalion. Just how accurate those figures were, I wouldn't know. After I left the first aid tent, a medical officer arrived and tagged everyone with frostbite for evacuation to Japan. All Marines treated by the corpsman, prior to his arrival, lost out on the Japanese evacuation. I don't know whether any Marines tagged with frostbite were counted in the muster figures I heard.

The following day, December 11, 1950, we were taken to the docks and went aboard the USNS Daniel I. Sultan. Our stay in North Korea had finally come to an end. All rumors of being home for Christmas were soon forgotten. Actually, with the tremendous large numbers of Chinese soldiers trying to annihilate us, it was a joyous feeling knowing they gave us their best effort, but fell short of destroying the 1st Marine Division. We came out with most of our equipment, including all of our wounded and a good number of those that gave their lives. This campaign would go down in history as one of the most memorable Marine battles ever fought against overwhelming manpower odds, plus frigid temperatures. As a survivor, I played a minor role in the breakout, but I will always remember cheating death or injury, while others less fortunate went home with serious wounds or in body bags. The Battle of Chosin Reservoir will not be forgotten.

Destination Pusan

It felt good to be aboard ship again, especially since our destination was Pusan, Korea. The most notable difference was the warmth of seamen aboard this ship. They couldn't do enough for us, going out of their way to make us comfortable. Ship's stores were opened up and what couldn't be found in the store's supply was offered by the sailors from their own personal possessions.

Talk about luxury. Taking a shower for the first time since October and it was now December, made me feel like I had died and gone to heaven. It felt good to shave and feel clean again. I still had the same old dirty rags I had been wearing for months, but that would change once we landed in Pusan. I now enjoyed the comfort of a warm haven. It almost made me forget how I had suffered, trying to stay warm without freezing to death. Ah, yes. I almost forgot about food. When was the last time I had eaten? It had to be in Hagaru, at least four days ago, with Tootsie Rolls to nourish me since then. The last real hot meal I had was our Thanksgiving turkey dinner, back in November. Navy chow made me forget my deprivation and starvation.

When the ship left port on December 13th, I was on deck making sure I had my last look at the land that had almost become my burial ground, as it had for so many others. It was a journey with memories I shall never forget. We arrived at our destination: Pusan, Korea, on December 14th. Then it was on to Masan, the new home of the First Marine Division.

Masan: A City within a City

On arriving at Pusan, trucks were waiting to take us to Masan, where tents were set up and ready for troops to occupy. The first notable change was the climate. It was a great deal warmer and there was no snow to remind one of the cold weather we left behind.

Beating the Odds

One of the first rumors making the rounds after we settled in had to do with our Air Force. I heard a Marine telling a story, concerning our Air Force wagering against Marines, being able to break out of the Chinese encirclement with odds as high as ten to one. It couldn't be done. Boy! Were they wrong! They probably weren't aware how the 1st Provisional Marine Brigade had distinguished itself in the Pusan perimeter defense. The unit had arrived during a critical North Korean offensive to annihilate United Nation troops, fighting desperately to maintain its toe-hold in a rapidly shrinking perimeter. The Brigade was immediately committed to combat in an effort to halt the steamrolling enemy attack. In its initial engagement, the Brigade met the advancing North Koreans, stopping them dead in their tracks. At the same time, inflicting heavy punishment on the advancing troops, Marine close air support, working with a proven system, aided the attacking ground Marines in winning the battle; forcing the enemy to withdraw. This was the first major setback for the North Koreans. Once again, Marine fighting ability played a crucial part in securing the Pusan Perimeter. Following this initial victory, the unit was used as a fire brigade, being shifted from one battle area to another to stem the tide of enemy aggression. Succeeding in beating the enemy in every confrontation involved going against the wishes of many Army commanders, who came to rely on Marine ability to defeat the enemy.

General MacArthur's command had originally requested a Marine division to add to its manpower needs in Korea. However, the 1st Marine Division, based in California when the request arrived, was at peacetime, low in manpower. A Marine brigade was quickly assembled from within the ranks of the division and expeditiously shipped to Pusan, Korea, while the remainder of the division hustled to reform its ranks to wartime complement. Upon completion of refurbishing, the division immediately set sail for Japan.

Most of the Division was still at sea while plans were being formulated for an amphibious assault to break the North Korean pressure on Pusan. The division was chosen to spearhead this historic landing. With its withdrawal from Pusan, the 1st Provisional Marine Brigade was to be returned to its parent division and rejoin the division at sea as the 5th Marine Regiment, joining the 1st and 7th Regiments that would make the amphibious assault scheduled to take place on the 15th of September 1950, at Inchon, Korea.

This was another example of Marine capability, with a successful landing, resulting in the seizure of Seoul and the recapture of all of South Korea. The Air Force should have known better in betting against winners, who had proven themselves since their arrival in Korea.

Tempus Fugit

Luxurious living became our lifestyle at Masan. There was no more crawling into foxholes at night or sleeping on frozen ground. Cots were used to provide bedding in our tents, plus we had the comfort of lighting at night. Our meager rations and hardships up north were soon forgotten with body building meals served daily. Rest and relaxation, plus physical exercise, aided in the recovery of health and morale of the Chosin survivors. Many of them were still suffering from their cold weather exposure. New replacements began to arrive, filling vacancies suffered by Marine units in North Korea.

Tempus fugit. It was Christmas, with church services to commemorate the Holy Day followed by a traditional holiday meal. Mail began to arrive with news and packages from home. It was sad to see a good percentage returned home or rerouted to naval hospitals. We had taken a manpower beating, with only a small percentage of the originals left from the Inchon Landing. Hopefully, more were in naval hospitals than buried up north.

New Year's Eve came with a bang. We began to receive our first beer ration, two cans per man. Wouldn't you know it; trouble and alcohol go hand in hand. That's just what happened. A New Year's Eve celebration occurred shortly before the New Year arrived. While resting on my cot in my tent, a rifle shot echoed in the immediate area, causing everyone to make a mad dash outdoors. I grabbed my rifle and followed.

Outdoors everyone milled around questioning the source of gunfire. There were no foxholes to offer shelter. It looked like the whole battalion was mustered, as they stood besides the tents, ready to handle any hostile activity. Finally, the mystery of gunfire was solved, with word that a Marine in a rifle company down the line, was the culprit responsible for the incident. With this information, tension subsided and Marines slowly started to trek back to their tents, resuming normal activities that were interrupted, when the weapon was discharged. I returned to my tent, lay down on my cot, and got lost in more memorable memories that had taken place the year before.

My carrier (the Leyte) was on a Mediterranean cruise at the time. There were options of leaves to Paris, France; Geneva, Switzerland; or Rome, Italy. I opted for Rome, and was granted leave that would take me to Rome for the Christmas Holidays. The highlight of this trip was participation in a Christmas Eve mass held by the Pope. On New Year's Eve I was on shore leave in Naples, Italy. Spending the earlier part of the afternoon touring the city, I finally ended up in a nightclub prior to midnight. We were having food and drinks, awaiting the arrival of the New Year. When the clock struck twelve, the town literally exploded with fireworks, the likes of which I had never seen before. It wasn't safe to walk the streets as spent debris fell from the sky. Now, this was a New Year's I'll never forget! And 1950 wasn't a total loss for me. A day or two into the New Year, promotion certificates were issued and I received one saying I was promoted to sergeant as of December 31, 1950. Three years/three stripes. I guess war casualties were creating openings for more NCOs and I lucked out.

Defective Ammo

We lost a great deal of our equipment up north, but managed to cheat the Chinese from their effort for total destruction of all our motorized vehicles. As an example, in our motor pool now was a beat-up jeep ambulance, causing all sorts of inquiries as to how it possibly could have gotten here. Just looking at perforated bullet holes covering the vehicle, raised a question of how anyone inside could possibly have escaped being killed. Yet, there was a chaplain by the name of Griffin, praying over a wounded Marine inside, when the ambulance was saturated with enemy small arms fire, killing some of the wounded. The chaplain's life was spared; when another Marine sacrificed his life by placing his body in front of the chaplain's to shield him from killing enemy fire. The chaplain was wounded in this encounter, but he did survive.

During our ordeal in North Korea, many Marines experienced a serious problem with weapon failure. Now was an excellent time to address the issue. A man's life depended on his weapon working properly. One day, we were taken to a secluded area that was setup as a rifle range, with everyone having a chance to test fire his weapon. Once the firing commenced, I was surprised at the number of misfires that occurred. I couldn't believe what was taking place. The problem wasn't with weapons; it dealt basically with ammunition we had to use. What occurred was an intermittent failure with defective ammo. I heard someone say it was old or bad stock and prolonged cold weather exposure may have added to our ammunition problems. That had to be the answer.

While firing my weapon along with others on the firing line, I was one of many experiencing problems with shells that were duds and wouldn't fire. How many Marines engaged in firefights with Chinese up north became casualties as a result of this faulty ammo? How many died would never be known. The exception would be the ones wounded that survived and were now in naval hospitals. They weren't as lucky as the two Marines in my platoon, who cheated death when their carbines malfunctioned, and the Chinese soldier had the same problem. Our new replacements were issued new rifles and now had an opportunity to zero in their weapons.

Now with each man receiving a beer ration, problems followed. For example, after drinking a few beers a Marine assigned to our tent became very emotional, took his carbine and went from bunk to bunk begging for someone to shoot him in the arm or leg. He had had enough of war, wanted out, and was willing to accept this kind of injury, inflicted by someone other than himself. With someone else pulling the trigger, he would have a perfect alibi as to how the wound occurred. No one volunteered to satisfy his wish. Finally giving up his effort, he crawled into his bunk and fell asleep.

The Chosen Few

Major Lawrence had been our company commander when we landed at Inchon. Captain Givens was next in line to assume command, and now Major Metzler took over as commanding officer of Weapons Company. He introduced himself with a speech I shall always remember. He said we were the chosen few among the rotten many by being here in this war, saying it may have been our brothers or fathers who were the chosen ones before us during the other wars, but now it was our turn. I had heard the saying once before while in high school by Mr. Doherty, a science teacher. It was his way of urging us to excel in our study.

Toward the end of our stay in Masan, there was liberty in town for those that wanted to visit the city. I was a squad leader. In my unit were two Marines that, in my opinion, had no fear. One we called Alabama, named after the state he came from; his real name was Higgins. He missed the activity up north, while assigned to a unit unloading supplies at the docks in the rear. He wanted to see combat, so he volunteered for duty with a rifle company. His request was granted, but he landed in Weapons Company instead. The other was called Stan Cholsky. He was also a new replacement. It didn't take long for the two to become the best of friends.

One night after returning from liberty, they stood by one of their bunks and began to unload their pockets. Falling onto the bunk were wads of Korean currency. Those in the tent watched and someone asked if they had robbed a bank. To answer everyone's curiosity, they pulled out M.P. armbands, saying they had waited until it got dark and then they placed the MP bands on their arms and raided the whorehouses, taking all the money that was available. It was surprising to hear that no one had challenged them during their scam. I can still picture the two as they sat gloating and dividing their night's take. What an ingenious pair.

Moving into Combat

Time passed quickly. Rumors of moving into combat usually started to circulate, prior to the actual relocation. Sure enough, a day or two later we were ready to move north. On the morning of our departure the company was formed for inspection. Roll call was taken to see if everyone was present. Then a Marine officer, with a young Korean adult, walked the ranks checking each Marine in detail as they walked by. This led to all sorts of speculation as to why or what this was all about. After the troops were dismissed, the mystery was solved.

They were looking to find someone who had decked the Korean the night before, causing some physical damage to his facial features. Needless to say, he couldn't be found. The Marine they were trying to find was the one who begged for a quick out with a shot in the arm or leg. He was safely hidden from view until the Korean was out of sight, returning to join us as we boarded trucks to move out.

Pohang Guerrilla Hunt

During our stay at Masan, our platoon was supplemented with light machine guns in addition to our other basic weapons. There hadn't been a great demand for flame throwers or rocket launchers, since our landing at Inchon. Machine guns would increase the firepower needs if requested by rifle companies. Our destination was somewhere near Pohang. This would be our base of operation, giving our new replacements a chance to get some combat training, while searching for guerrillas operating in this area.

One day, I was ordered to take my squad up to a certain location on a ridge with heavy machine guns located below us on my left flank. I had one light machine gun that was immediately set up, facing open terrain covering the approach to our position on the hill. While we sat and chewed the fat, some suspicious activity was observed taking place to our far left in the area where the heavy machine guns were located. We could see individuals moving down the hill, dressed in green with what looked like dangling ammo pouches hanging from their shoulders. If they were guerrillas, why didn't the heavy machine gun open fire?

I had no field glasses or radio, so I sent one of my new replacements, Dale Erickson, to run over and check with the machine gunners on what was happening. As quickly as he left, that's how fast he came back yelling, "Gooks, gooks!" At this time my gunner was ready and waiting for orders. When I yelled to open fire, everyone else joined in with tracers, setting the hill on fire. Almost immediately the machine gun jammed. The guerrillas disappeared from view before the gun stoppage was cleared.

I had no communication with my platoon leader or the company. I had to get word to them as quickly as possible that we had encountered some guerrillas and needed to know if they wanted to pursue them. That wasn't necessary. The sound of our firing had a company officer coming up to investigate the ruckus. I explained to him what had taken place and asked whether he wanted to chase the guerrillas. He told me to stay where I was and he went back, returning with the rest of the company. The chase was on.

The brush we set afire was still burning. We looked around, without finding any bodies or trails of blood. Based on the lack of any blood, we may have just scared the daylights out of them. We then raced down the hill, trying to find the guerrillas without success. We finally gave up the search and returned to bed down for the night.

A puzzling question remained to be answered. After our return back to camp, I asked the company officer who was the first to appear on the scene, why the heavy machine gunners had failed to open fire at the guerrillas. He responded by saying the enemy was on top of them before they knew it. I then asked why they held their fire after the guerrillas slipped by. The reply was that they were too scared. Even though most were new replacements, they should have shown some fighting spirit. This wasn't just another training exercise. To the enemy and members of my unit, this was a life and death encounter. My unit was alert and reacted immediately on confirmation it was guerrillas moving down the hill. The heavy machine gun crew had a golden opportunity to take some enemy soldiers out of action and failed to do so. There was no excuse for this gun crew to freeze up in their initial encounter with the enemy. How could anyone rely on their fighting support in the future?

After this activity, I was given charge of Sergeant Lister's old section. Most of our time spent in the hills equated to conditioning hikes, with no enemy spotted. One day, I had my section search for guerrillas in a hilly area where an occasional rifle shot could be heard echoing in the distance. We were almost to the top of a ridge, when we came across bodies of enemy dead who had been left to rot. From the looks of them, they had been there a good while.

With our movement up the hill, sporadic gunfire could be heard at intervals. It sounded far away. Approaching a clearing, some Marines started to bunch up. Another shot report echoed from a distance and that's when one of my guys yelled. He was hit. A sniper bullet had pierced his arm. We were lucky to escape any other seriously wounded. The corpsman applied first aid and we continued to the top of the hill and down the other side keeping our regular interval and offering less of a target, if more sniper fire came our way. Our wounded man was a new replacement and my first casualty. His war probably ended for him, depending on how seriously he was wounded. That shot bothered me, not knowing if it could have been fired by one of the guerrillas that eluded our chase a few days back.

Our guerrilla chasing days had to come to an end, sooner or later. It was an excellent training exercise while it lasted, testing one's physical condition. It gave the men a chance to experience vigors that they would face shortly, with those unable to keep up with physical demands weeded out. Many who had been in a soft position at their former duty stations experienced difficulties in keeping pace. As time passed, their bodies accepted the physical demands made on them in Korea.

I recall a Marine that was scheduled for discharge in May of 1950, but because of a medical problem was hospitalized until the Korean War erupted. Harry Truman's one year extended tour of duty negated his discharge, so upon leaving the hospital he was assigned to Weapons Company and became a member of our section. When he joined our unit, he was the largest obese individual in the company, needing a size 46 to 50 or larger belt to hold his pants in place. Yet, from the time of our landing at Inchon, this Marine started to shed weight from the rugged physical demands experienced on a daily basis, until our arrival at Masan. When told what this man had weighed when he landed at Inchon, the new replacements couldn't believe he once carried over 300 pounds on his slender physique. He now looked trim like all the other Marines. He lost so much weight that a size 32 pair of trousers fit just right. This was an example of how physical demands on his body, plus field rations, conditioned him to lose all that flab in a span of three months.

Fast-rising Water

Once again we boarded trucks, heading north to where the action was located. At night, we stopped to bed down in a dry riverbed. Trucks were emptied as Marines set about to bed down on the ground for the night. That's when my truck driver offered me his cot to use, if I wanted to accept. I didn't want to deprive him of his comfort, so I asked him where he was going to bed down. He told me not to worry; he would be just fine sleeping inside his truck. It was an offer I couldn't refuse, so I took the cot and made my bunk alongside the truck. It started to drizzle so I rigged my poncho to protect me from the rain. I fell asleep, only to be awakened by a great commotion taking place around me. What had been a dry riverbed when we stopped to bivouac was now a flowing stream. Everyone lying on the ground found themselves being flooded by the rapidly rising water. It was a gloomy night with heavy rain coming down, forcing everyone to move personal gear to high ground outside the riverbed. I was on the bank breaking down my cot, when our trucks started to move out to higher ground.

Not everyone succeeded in recovering all their equipment. After taking inventory of my personal gear, I had to do a double take. I had my pistol and cartridge belt, but my rifle was missing. Did some other Marine grab it by mistake or was in resting at the bottom of the stream where we had stopped to bivouac? I walked over to the water's edge with intent to enter the stream to recover my weapon. I couldn't believe how quickly the water level had risen. It was now impossible to enter the stream.

Trucks moving out of the water to escape being caught in a quagmire may have run it over, burying it deeper in the mud. Any optimism I had of finding my rifle vanished. Another deterrent was the cold and darkness. Hopefully by morning, the rain would stop and then I expected to be able to enter the stream and find my weapon. Nature had other ideas, however, because when morning came, the rain was still coming down hard. The river had swollen with a current, making entry into the water impossible. A chill in the air kept troops moving to stay warm. I still had my pistol, but a rifle was needed for added protection. A pistol was only good at close range, while a rifle gave one added protection at a greater distance.

Inventory was taken to see what equipment was lost during the night, so it could be replaced as soon as possible. I felt ill at ease as word came to climb aboard the trucks. It was time to move out. Up until now, I was traveling inside the cab with the driver. Once inside, I told the driver of my sad misfortune that had occurred during the night. He said he had an M1 rifle that would be more useful to me than to him. It was mine, if I wanted it. How lucky could I be? His generosity lifted my spirits as I gratefully accepted his offer. I felt more secure knowing I didn't have to worry about securing a replacement for the one I had lost.

We moved north all morning, coming to a bridge made of wooden logs that appeared to be almost totally submerged. The convoy stopped while Army engineers worked to make the bridge passable. This gave us a chance to exit the trucks, with everyone attracted to the work in progress. The rain finally stopped. It was now a challenge for engineers at work to prevent logs from being washed away. It looked like the logs acted as a dam with water cascading over the top. After considerable time had elapsed, we got back aboard our trucks and managed to cross the stream without mishap.

End of Vacation

It was now almost the end of February and we were going back to face the enemy. Our vacation was over. Late in the afternoon we moved into position on the line. Our perimeter had us facing a stretch of flat open ground as we dug in for the night. Shells could be heard whistling overhead. As I sat in my foxhole, I could see an explosion occur roughly a hundred yards to my immediate front. When we first got here, I was told there was a minefield located in front of our perimeter. This alerted everyone for a possible attack, but none came. During the night, when I checked my machine gun position, I found the gunner asleep. He was a new replacement who was startled awake from the violent boot that popped his eyes open. He was told there wouldn't be a next time, if caught asleep on watch again.

When it became light, I could see bits and pieces of what was once a Korean cow. It must have stepped on a mine and blown itself apart. We moved onto some high ground the next day and dug in for the night. In the morning, some Marines were found bayoneted in their sleeping bags. They paid with their lives for failure to keep alert. Hopefully this information sunk in with our new replacements, especially with my gunner after the incident the night before.

The following night, the temperature dropped. When morning came and I took my shoepacs off to change my socks, I was shocked to see both of my feet had turned black. Was this another frostbite occurrence? I told my platoon leader of my problem. He directed me to report to a medical aid station a few miles in the rear. Some Marines were wounded during the night, and since I was going back to the aid station, I was detailed as a stretcher bearer to carry one of the wounded. There were four assigned to a stretcher.

Since my arrival in Korea and all my exposure to deadly enemy action, I had never been tagged to a stretcher detail. I would now experience what others endured in this assignment. At first, it was no problem, but very quickly it seemed like our casualty weighed a ton. It was a grueling task bringing this Marine down off the hill along a narrow pathway. I struggled all the way to maintain my balance and not lose my hold on the handle. We finally made it to the aid tent, where our wounded man was given immediate attention. I then reported my problem to a corpsman. He told me to take my shoepacs off and had me insert my feet into a bucket of water. After a long, torturous trip down the hill, my normal color had returned so the corpsman said that my feet looked okay. I was then told to return to my unit.

Decimated Army

We were into March now and on the move again. As we moved forward into a partially wooded field, I could see a dead body laying face down on the ground. There were more bodies scattered in the wooded area. A closer look at their uniforms identified them to be Army soldiers, but what outfit were they from? "My God," I thought. "What had happened here?" The further we walked, more dead bodies were seen lying everywhere.

We walked onto a road and coming to a bend what I saw up ahead was beyond belief. An Army tank had pinned a Chinese soldier against a cliff. He was pinned waist high with both of his arms extended forward like he was trying to stop the tank. I can't imagine how he could have gotten himself into such a position.

Moving further along the road, we came across all kinds of vehicles. Some were destroyed; others just abandoned. We came to an area where a number of jeeps were disabled off the road. One jeep that drew the most attention, had a driver slumped over the steering wheel with a screwdriver sticking out of his head. To the sides of the road in the field were the bodies of many American soldiers, most of whom were completely stripped of their shoes and clothing, while others still had some underwear remaining. The Chinese had taken everything they could.

Word came down that a few Army survivors had been found, having escaped being killed by finding shelter in a cave. They were attached to an Army artillery battalion moving in this valley, without any hillside protection. Chinese soldiers had caught them without flankers and had had themselves a turkey shoot. The entire Army unit was decimated, with all useable field guns captured by the enemy.

Marines checked out some of the vehicles that looked okay. Those vehicles now became Marine property. A Marine entered the tank that had pinned the Chink, and it started up with no problem. Seeing this massive carnage, one Marine in our platoon reached his breaking point. He went into a crying jag and had to be evacuated. After seeing the massive killing and being told that the Chinese were just up the road waiting for us, another Marine got so scared he wet his pants. No one dared to make fun of his accident, as he tried to explain how it happened. Some of his buddies tried to reassure him that he wasn't the first Marine to experience an involuntary bowel movement during trying combat conditions.

Dale Erickson came over saying how deeply he was affected by this carnage and asking if he could possibly be transferred to a rear unit. He couldn't bear to look at all the killing that had taken place. I told him encounters with the enemy in battle resulted in what he had just seen and he would be okay in a day or two. I don't know whether he believed me or not, as he walked away.

Holy Picture

We moved further up the road and stopped. There was an Army vehicle a short distance from the road, so I walked over and came to a pile of burned rubbish located a yard or two from it. My eyes were immediately attracted to what lay on top of the ashes. There I saw a holy picture of Jesus that had miraculously survived the fire. I reached in and pulled the photo out of the rubble, examining it closely. It appeared to have a slight watermark and must have come with an Easter card. The rest of it was gone in the fire. With all the vehicles that lay battered and disabled, why had I been attracted to this one? Was there a reason why I was drawn to that particular spot? I was so engrossed in my find that I completely forgot to check the truck for other worthwhile possessions. I walked away with the icon, questioning how it had survived all this carnage and deciding I had best take possession of it and maybe I, too, would share in its survival.

Tough Going

That night, I was given a position to occupy on the high ground overlooking the disaster area. Artillery shells were landing in the area, exploding short of the ridge but near the road below. In the morning, I took my section down the hill and came to where the shells had exploded during the night. The ground was covered with huge chunks of shrapnel, the size of steel that could only have come from a large sized, bored weapon. I could just picture the damage it would have done had we stayed in that area during the night.

We moved forward during the day and I moved my unit onto a ridge, digging in on the reverse slope of our objective for the night. Enemy artillery fire kept hitting the forward slope or going over our head. We all questioned whether recently captured Army artillery pieces were now responsible for all the shelling coming our way. There were no casualties from enemy shells that night.

During this cold spell, I came across a new, abandoned parka. Although mine was battle scarred but still serviceable, I took possession of it and carried it as a spare in my waterproof bag that I had strapped to a new type of "A" frame recently issued. Hauling the extra weight was an additional burden, but the payback in using the parka to wrap around my sleeping bag at night was worth the price I paid in return for the added warmth and comfort it gave me.

We moved higher into the hills, away from the roads. Climbing one hill took a toll on the troops hauling full equipment. It was a snail's pace climb all day with many halts until word came to bed down for the night. We still had a long way to go to reach the top. The journey continued the next day until we reached the peak, only to see more rugged hills waiting to test our resolve.

Resupply posed a problem. Korean laborers were used to haul food and water in five gallon cans up to our positions, plus ammo if needed. Another problem arose, dealing with how to bring up .75mm recoilless rifles. This was a heavily wooded area that our operational officer felt needed these guns for defensive positions to help repel enemy attacks should they occur. I don't know if that problem was ever resolved.

Dignitary

A key town was taken in our sector, resulting in my section moving down to a road, leading into the newly-taken prize. On the road, we dropped our packs and made ourselves comfortable, while waiting for word to move out. We had been there a short time when it looked like some jeeps were heading our way. They were approaching rapidly. As they got closer, we could see they were Army M.P. jeeps with standing occupants on the passenger side of the jeeps, dressed in clean khaki uniforms. They were holding unto the windshield and yelling for us to move off the road. We wondered what could be so demanding in this area of the battlefield to warrant obedience to their request. The answer came with the occupant in the last jeep. They were clearing the road for General MacArthur, who rode in that jeep. We were still resting, awaiting word to move out, when the jeeps that came by earlier returned, exercising the same procedure used earlier to clear the road. With his mission accomplished, whatever it was, General MacArthur was now sitting in his jeep holding a corncob pipe to his mouth, as he again passed by. I had never seen the General in person before, but that day I finally got to see the man responsible for me being there.

Combat Breeds Death

The following day we moved out, crossing some rice paddies with irrigation trenches filled with water. Every now and then, we came across the remains of Chinese soldiers lying beneath the crystal clear water. I don't know whether the soldiers died where they laid or if they were placed in the trenches to be buried.

We arrived at a stream. There was a bridge located a short distance from where we stopped. Some Marines ambled down to replenish their water supply, filling their canteens, using purification pills to make it safe to drink. We walked a few yards up stream, past the bridge and discovered a mass of dead Chinese bodies lying in the water. I looked around, but saw no one dumping out the contents of his canteen.

Later that day, Dale Erickson came over and again questioned me about getting a transfer to a rear unit. Dale was a good Marine and had showed some grit during, our guerrilla chasing days. I didn't want to lose him. He was a gangling kid, just eighteen years old with bad eyesight, and requiring glasses to see properly. In all the time I spent with Weapons Company, I had never heard of anyone asking for a transfer out. Usually, it was the other way around; with rear echelon Marines wanting transfers to a line company. Since his eyesight was bad and he wore glasses, one of his friends made a suggestion that maybe, if he had a problem with his glasses, he might just have his wish granted. A day or two later it happened. Dale came to me showing me his broken lenses. I had no recourse but to send him back to the company CP to handle his problem. I said my farewell to him as he left for the rear, never expecting to see him again.

I miscalculated on him ever returning, because after a week or more had gone by, Dale returned with a brand new pair of glasses. When I asked him what happened, he gave a rundown of events that took place after he left the section. He was taken to a hospital ship, named Repose to be fitted for new glasses. Once aboard ship the medical officer, a psychiatrist who examined him, decided he needed more than just glasses. They kept him aboard a few extra days, where he was well fed and rested. At least he had some R&R, which was more than we had.

After his return, he never again raised the issue of a transfer to a rear area.

We had been in the front line for a good period of time now, where shelling by the enemy kept everyone on edge. One morning, while resting on a hilltop, someone yelled, "Incoming." Everyone took cover, only to discover that a Marine had just been brushing his teeth and the noise heard was mistaken for incoming shells.

Combat breeds death and destruction. Our units were no exception. Deaths and wounds were decimating members in our companies and there was a need to replace those missing. I still have this vivid memory of new replacements being brought up to our line positions, while we were heavily engaged. Enemy mortars and artillery were working over our defensive positions, as new arrivals were replacing casualties. The replacements were getting wounded or killed on their first day in combat, never having a chance to leave any lasting impression on others. They were just more names added to our casualty list.

During this day of heavy shelling, two of my Marines went to make a nature call. Shortly thereafter, one of them ran back, yelling that a mortar shell had wounded his buddy. It exploded, while he was taking a crap and shrapnel ripped into his stomach. What an eerie feeling that must have been, getting wounded with his pants down, attending to natures need.

The next day, enemy artillery shells caught us crossing an open field. Everyone took cover until the shelling eased. They were trying to stop our advance. We moved out of the area before anyone got hurt. We came to a road. My platoon leader assigned my section to help engineers clear the road of any enemy mines. My section provided protection while the engineers moved in front of us, probing with metal detectors as they searched for planted mines. We spent a good portion of the day doing this, not knowing if the enemy was waiting to pick us off.

Finally, the engineers completed their mission and I was told to return to my unit that was now located on one of the high ridges above the road we had just cleared. I found my company and went to see my platoon leader, who was involved in a conversation with the mortar lieutenant. Just then, enemy mortars started to rain down. My platoon leader hit the deck, landing into a pile of fresh human waste. The mortar officer was Lieutenant Henderson, who had been wounded at Hagaru and had recently rejoined the company. I watched as he took out his map, checked his location, and immediately issued orders to one of his mortar sections. The guns set up to his coordinates and fired their shells. The enemy fire ceased with no more shells falling in our area. I asked him how he did it and he said it was pure luck. Maybe the platoon leader falling in the honey pile brought us luck.

My platoon leader was a replacement for Lieutenant Kavagavich. This new officer, whose name I can't recall, suffered from wounds he had received at the Chosin Reservoir and still carried aches and pains from that encounter. He was "A Okay" in my book. The night before our departure to the front lines, he called me into his tent, brought out a bottle of booze, and told me to help myself. That was a good way to ease the tension, and the liquor was great.

The next day, we moved into a valley and set up a perimeter on a small hill. The Lieutenant came over and told me that our line companies were going to withdraw through my position at night. I then asked him if I was to follow after they passed through. I couldn't believe his reply to me. "No," he said. "You will stay here and die, but you won't pull back." He then asked me what additional supplies I needed. I told him at least five cases of grenades and a lot more machine gun ammo. Shortly thereafter, a jeep arrived with my supplies. I then distributed the ammo so everyone had a fair share. Night came and we waited for the rifle companies to file through our area. Nothing happened. No one approached my area during the night. In the morning, I asked the Lieutenant what happened. He said there was a change in orders. The pressure from the Chinese never materialized; therefore, there was no need to pull back. When word finally came to move out, all the additional ammo in our possession had to be returned.

Worthless Army Boots

For whatever reason, the 7th Marine Regiment was chosen to relieve the 8th Army Regiment of the 1st Cavalry Division in the line. We arrived in the Army sector to make the relief. About that time, I was sick of my leggings, so as we moved into the Army area, it seemed like the perfect opportunity to acquire a pair of Army boots with the high straps that would eliminate a need to use leggings. I inquired where the Army shoe supplies were located. Once inside the tent, I asked the supply sergeant if he could spare a pair of Army boots. He said, "No problem, help yourself," and showed me where the boots were stored. I found a pair to fit, walking away like a peacock, because I now had a pair of boots where I could stuff my lower pants inside the boots and buckle up. Leaving my old boondocks behind, I felt that I had made a good trade. But that was not the case. Within a short period of time climbing rugged hills, the heel on my right boot came apart and shortly after this occurred, we were ordered to withdraw to the rear.

My section took up a temporary position on a hill, overlooking a road. We set up our machine guns and waited while other units passed through our area. As the Marine artillery battery moved by, I saw a former shipmate, Charley Chapman, who I knew from my duty aboard the carrier U.S.S. Leyte, but there was no time to get reacquainted.

Finally, it was our turn to move. During this march to the rear, it started to rain. That's when I lost my other heel and ended up marching close to twenty miles without heels. I couldn't wait to acquire a pair of reliable Marine boondockers. I got those almost immediately, and from that time on, I learned my lesson and had no further interest in owning any Army boots.

Chinese Casualties

Having accomplished our mission while attached to 1st Cavalry Division, detach orders arrived, sending our regiment back to rejoin our parent division. We were again involved in moving forward against the enemy. I found myself walking on a road heading north. Marine artillery and mortars paved the way. Chinese soldiers were now surrendering in groups or alone. It was an eerie feeling walking the road, only to have a Chinaman pop out of the ground in front of us with his hands up. While on this road, I saw a Chinese unit to my right move in behind a hill. I didn't know if they gave up or had to be taken out.

We came to a heavily wooded hill and were told to secure the slopes. Artillery fire peppered the ridge prior to our assault. I watched as the shells blew away trees and limbs. When the shelling ceased, we started our move upward. There was no enemy fire coming our way, but the hill was covered with dead Chinese. The artillery fire made a shambles of the wooded area we were now scaling. In this one area, I was attracted to a dead Chink, who was in a kneeling position in his foxhole. He was still holding onto his rifle but half of his head had been blown away. He had probably been dead only a short time, as brain fluid was still visible in the cavity that remained.

We got to the top and were told to dig in. I had my machine guns set up and waited for the next events to take place. It started to drizzle. During the night, there was some movement to our front, so a machine gun opened fire. The next morning, I walked over to the area where activity had taken place during the night and found a dead Chinese soldier. He may have been with others, but he was the only one to die. Up until then, I had never searched any enemy dead, but I made an exception in this case and went through his belongings, finding some Chinese and Korean currency. One Chinese bill had a 500,000 dollar figure on it; there was a Chinese or Korean bond; Korean 100 Won notes; plus other Chinese currency. This soldier also had a long sock wrapped around his shoulder that contained his provisions.

Soaked Through and Through

Word came to move out. There were some Marine casualties during the night that had to be moved off the hill. It became our assignment to make sure they got to the bottom of the ridge, where medics were waiting to tend to their needs. With this chore completed, we moved out. After a while it started to rain, so I covered myself with my poncho for protection from the rain. The rain wouldn't let up and although I had rain protection, it didn't help. The longer we marched, the wetter I got. The rain caused the poncho to sweat on the inside, and before long, I was completely saturated in wet clothing. I wasn't alone in experiencing a complete soaking. There was no respite from the misery the rain created. I tried to decide which was worse, the rain, or the snow up north. I could feel the cold wetness against my body. I felt like a sponge retaining all the water in my clothing, with water draining down to my shoes, keeping my feet cool as I walked. The rain finally stopped. It felt great to be able to build a fire and let the clothing dry.

I carried a rubberized sack that now proved to be a real blessing. All my possessions inside were spared from getting wet, including my sleeping bag. A few days back, when the weather started to warm up, everyone was told to surrender his parka and winter clothing. Boy, did I miss those parkas now. A new item that found its way to our unit was a supply of air mattresses. The mattress gave extra comfort in the foxhole, but required additional time to inflate each time the owner decided to use it.

With all the recent rain we had, all sorts of problems were created. Sections of roads and trails were gone. It was surprising how Korean natives appeared out of nowhere, working together to rebuild the damaged surfaces. They scoured the immediate area, searching for stones or rocks, dumped their loads, filling the muddy holes until it was safe for vehicles to pass.

Operation Mousetrap

We were operating across the 38th parallel. My company moved into a forward position and I was given an assigned area to defend. In the process of digging our foxholes, the Lieutenant came over and told me my section had to be relocated. We were taken quite a distance from the main company to a road, crossed a bridge over a river, turned left onto a road running alongside the river, moved a short distance, and finally halted.

84

I was to set up a defensive position, covering all approaches to this area on a small knoll, located adjacent to the right side of the road. I went about assigning individuals to defensive positions and told a couple of my men to take up positions on the other side of the stream to cover the road and stream area on our left flank. There was a steep bank on the other side, and halfway up was where they decided to dig in. It wasn't long before I could hear them yelling. They had found a body where they were digging. It seemed like an odd place to bury someone on a riverbed.

As I looked over the area we were assigned, there was no one anywhere near us. The only people in the vicinity were an Army artillery unit, and they were located at least a mile away. There we were with no communication link to anyone. We were completely isolated, not knowing what to expect.

Fortunately, after spending a couple of quiet days there, we were finally ordered to rejoin the company that was still in the same position as it had been when we left. While we were gone, South Korean battalion workers had moved in to occupy ground within our perimeter. Almost immediately, it became evident something was wrong. The Koreans had no sanitary discipline. They were like animals, leaving their waste uncovered everywhere inside the perimeter. The place stunk so bad, I found myself gagging from the stench and had to be careful where I stepped to avoid coming in contact with Korean crap. Thank God our stay there lasted only a day.

I was told that our mission on the knoll was part of Operation Mousetrap and that my section was the bait. The following day, the company moved out. We passed the knoll area and had gone a short distance when we came into a killing field. Chinese dead lay everywhere. They had been killed during the night. A bulldozer had dug a huge trench. Korean natives were taking the dead Chinese bodies and dumping them into the trench. My section had lucked out, by moving out a few hours prior. The Chinese troops moved into what became their graveyard. I never did learn who was responsible for this carnage.

We moved into another mountainous area. I was assigned a roadblock to defend. There were some Marines wounded during the night. In the morning, a helicopter landed a few feet in front of our roadblock to take the wounded out. With the chopper loaded, I watched the takeoff and then saw a disaster occur. The chopper dropped, smashing into the ground. The pilot was badly shaken. I heard him saying to an officer that he had hit an air pocket and lost control. He was lucky this accident occurred before the chopper had gained more altitude, otherwise, the wounded who were aboard might have all been killed with the pilot, instead of badly shaken. The chopper sustained damage that was beyond immediate repair. The officer in charge radioed for a truck to come and recover the wreck. Damage wasn't that bad and the chopper could be salvaged.

We started to move out, marching on a road adjacent to high hills. Someone said that we were being relieved and shortly trucks came and took us to the rear, dropping us near a river in Chunchon. Before anyone could make themselves comfortable, a Marine officer came over and told us to stay loose, stay in the area, not to wonder off, because we might just be going back into the line again. Sure enough, toward evening, the trucks arrived. We got aboard, and were again headed toward where the action was. It was fairly dark as we traveled along this road. We passed a column of trucks, but in the darkness it was hard to see the remains of an Army truck convoy that had been completely destroyed by the Chinese. The Army unit in this area ran, losing most of its equipment such as vehicles, mortars, machine guns, and personal weapons.

We arrived at our destination, only to find that Chinese soldiers were everywhere amongst us. Prisoners were being rounded up, as they roamed in the dark. We bagged a good number of them during the night. At dawn, we got to see where we were. I was surprised to see French soldiers as they moved out of our area to a new location. I moved my unit onto a hill and waited. Just then a terrible explosion took place. Someone hollered, "Incoming!" and everyone sought cover. At the bottom of the hill in a field, I could see a 4.2 mortar crew in action. Shortly, word reached us that one of the 4.2 guns firing had a shell that hit an overhead communication wire, causing the explosion that made us seek cover. It also created casualties among the gun crew. From our position on the hill, we had a bird's eye view of the gun crew casualties. Speaking of wires, I was amazed at the Chinese efficiency in recycling communication wires, left behind by retreating UN troops. Usually, the battlefield was stripped of all wires, when lost ground was retaken in a following attack.

Caution around Civilians

There was information making the rounds, concerning Germ warfare. A cholera epidemic had apparently played havoc within the Chinese army, with Korean civilians also infected. I recall moving into a village with all the natives staring at us. Many of the natives we passed had that sickly look about them, looking to us for medication. We had none to give them.

That night, we set up our defensive position on a hill above the village. Our machine guns were set up facing a hutch roughly a hundred yards to our front. We got there late and had no time to check out the hut to see if anyone occupied it. If an attack should occur during the night, it would likely be the direction it would come from. It turned out to be another quiet night with no activity, until the break of dawn, when a person came out of the house and made a slow walk toward us. It turned out to be an old mamasan carrying a wooden bowl in her hands. She wanted us to accept her small offering of beans that she had carried in her bowl. With all the sickness in the immediate area, no one was willing to accept her offer. Instead, almost everyone dug into their rations and buried her with food to last awhile. We then moved out, watching the mamasan smiling and bowing, not knowing whether there was appreciation or hate, within her. After all, this was North Korea, and we were the enemy.

We moved into an area alongside a stream and stopped for a break. While resting, an old Korean approached us begging for medicine. We could tell by the looks of him that he was sick. He looked very pale with teary eyes. He was told to see our medic, but the corpsman carried no medicine that would help him. Being rejected, the old Korean then sat down near the stream and cried like a baby. He wasn't ready to die. It was a sorrowful scene, but this was war and we weren't the ones responsible for the sickness he had picked up from the Chinese.

I looked around the landscape and noticed odd pieces of furniture lying on the ground in an open area, away from the huts. It looked like the civilians were planning on moving and were taking their furniture with them. We finally moved away from the stream area. Making a patrol around some huts, since this was North Korea, some lit cigarettes were discarded onto the straw roofs and it wasn't long before the huts started to blaze. Our officer quickly put a damper to the random arson.

We came to a village and started to check the huts. I entered one that had been fire bombed and found the remains of its occupants. The male of the household was barbecued, completely charred. I could tell it was a man by the remains of his genitals, but I couldn't tell whether he was a soldier or civilian.

Time for a Break

We had been in the line now for close to eighty days without relief and it was time for a break. Trucks came to take us back for a rehab rest. We arrived in Chunchon and located near a river. We set up our shelter halves and got ready to relax. There was hardly any time on the line for personal hygiene, such as showers, clean clothing, shaving or haircuts. Once our tents were set up, the first order of business was to clean up.

I walked over to the riverbank where Korean women gladly washed our clothes in exchange for cigarettes or candy. Others wanted American military currency. The weather was nice and warm, so all I had to do was take all my clothing off and hand them over to a Korean mamasan who then washed them, while we bathed or swam in the river. It seemed like no one felt uneasy, undressing, and walking around naked in front of the Korean women. Everyone was doing it, while his clothing was being washed.

It had rained before we got there, so the river level had risen. There was a metal bridge span a couple hundred yards away that was blown up and looked like a twisted pretzel. On entering the river, I found a very strong current that could carry a person swimming toward the blown up bridge. I waded out about halfway across, letting the current carry me downstream. I didn't realize how polluted the water was, until I saw human waste floating by. That's when I lost all interest in swimming and couldn't get out fast enough. By then, my clothes were washed and dried from the sun's rays.

With plenty of free time on our hands, John Kraus, Dale Erickson and I decided it was time for some R&R. We planned to start early the following morning, hitching a ride from an Army vehicle that took us all the way to Wonju. Since all of our time was spent on the line or reserve close to the front lines, we were looking forward to seeing what kind of activity took place in the rear.

In Wonju, we came across all sorts of Army units occupying most of the area around town. We set out looking for a place that offered entertainment, ending up in a restricted section of town, made up mostly of wooden shacks. There was a hooker, trying to sell her body. Army MP's riding in a jeep spotted us and they were making a turn in our direction, when the prostitute, upon seeing the MP's, took off. Since we were AWOL from our unit, we decided to follow, scaling a fence. Kraus experienced a problem, cutting his hand on the sharp wire, but we got away, losing the MP's.

We decided it was time to end our stay. We had no further interest in wasting our time in this godforsaken town. Before leaving, we had to be fed. Finding an Army mess tent was no problem. Once inside, the Army personnel asked how we had gotten there. After hearing our story of where we came from and why we were there, Army cooks supplied the mess gear and were more than happy to feed us. We got back before dark, playing hooky for the day with no one the wiser.

While still in reserve some good, morale boosting information was issued to all Marines. A rotation system was being put in place. Now there was hope of survival with something to look forward to. Prior to this release, death or being wounded was the only way out. Tennessee Toddy was one of the first to be notified that he was going home. All the old timers were happy for him. The war was over for him and he had survived.

Ingenuity is a Marine's trademark. While in this reserve area, someone came up with a novel idea. It provided a light in a pup tent for anyone interested in having one. All that was needed was an empty ration can filled with lighter fluid, a wick, and "presto" light was only a match away.

Back on the Line

Once again rumors of going back into combat made the rounds. A day or two later, it became official. When we got up early and went to chow, there was a surprise waiting for us. Steaks were the breakfast specialty. There is nothing like being well fed before going into combat. But, the best was yet to come. Once everyone was squared away to move out, each Marine received a quart carton of fresh milk to take with him, courtesy of the aircraft carrier Philippine Sea.

Our transportation came to take us to where we were going. On arrival at our assembly area, we pulled off the road, got off the trucks, and had started to move away from the vehicles to form up on the road, when suddenly there was a jarring explosion. After unloading the men that were aboard, one of the trucks started to pull out and ran over a land mine. There were a lot of lucky individuals spared injury or death, when the truck originally stopped short of the mine. Once again, this was war and the grim reaper waited at every turn. Death or being wounded was the quickest way out, other than rotation home.

We moved out and, coming to an open field, stopped while Marines with bayonets probed the ground for mines. We got the word to move out and crossed open terrain, not knowing if the next step would be our last. Back on the road again, we arrived at a hilly area. The platoon leader ordered my section to carry out an assault on the ridge. Once in position, we started to move up the slopes of the hill that had a heavy growth of brush and trees. I was expecting the worst, but there was no enemy to offer resistance.

Savoring the Beauty

In this land of death and destruction, there was also a time to savor the beauty of the terrain we traveled. I remember entering one area that had the most beautiful layout I had ever seen, where nature paid special attention in its design. It was just breathtaking to be able to enjoy the scenic layout. Even our digging failed to alter its aura. It was like living in a different time zone with no roads, electricity or plumbing. Almost all the roads were dirt covered or narrow pathways and the day ended when it got dark. Some nights, while occupying a high mountain ridge, it seemed like the moon would land on its peak. It was so close that it could almost be touched.

One morning, while waiting for our turn to move out, a Marine was seen carrying a small deer over his shoulders. I could just picture the feast the owner and his friends would have once the deer was butchered into steaks. The menu change would only last until all the game was gone, and then field rations would once again be the staple of diet for every meal. Once in a while, we flushed out a pheasant. At first, I wasn't sure what kind of bird it was. I had never seen a black pheasant before. But I was assured that's what they were.

Death at Any Moment

Up until now, Marines were always on the move with the enemy falling back. We got to the top of one objective and had a panoramic view of what awaited us in the near future. It was like a time warp. Looking at the succeeding ridges, it reminded me of the First World War. The Chinese had dug trenches around the hilltops. I had never seen this kind of defensive position since my arrival at Inchon. Would this hinder our success in action against them? Time would tell. The next morning, we moved down the hill, reached the road, and started to move out. A jeep passed us and moved to the front of the column where it ran over a land mine, killing the driver.

During our march, word came back that Dog Company was heavily engaged on a ridge up ahead. A new platoon leader from that company was killed in this attack. There was talk that flame throwers might be needed to dislodge the enemy. While waiting for specific orders, my company moved off the road, climbed a hill and occupied an area near a Chinese trench line. I now had a chance to inspect the construction of this defensive line. Overhead protection was erected every so many yards, but the trench itself wasn't as deep as I thought it would be. As I stood near the trench, an explosion took place and someone yelled, "Incoming!" That's all it took for everyone to seek shelter in the trench.

In the trench, we waited for more shells to drop, but there were no more explosions. Leaving our shelter, we started down to the other side of the hill. That's when the bad news hit us. Out of curiosity or maybe with intent to salvage some enemy supplies, a Marine sergeant had picked up a bag full of enemy grenades, slung it around his shoulder, and went on to meet his maker. That was the explosion causing everyone to seek cover. The bag must have been booby trapped. Bits and pieces of his body were strewn all over the hill. I can still picture a portion of his bare upper leg resting on the ground, with members of his unit trying to recover as much of the bits and pieces of his body that could be found. The need for flame throwers was forgotten.

The Chinese were offering stiff resistance, but our rifle companies were forcing them to withdraw. I was given a mission to accompany four tanks in an effort to open up a road. The rest of my company and battalion were involved in trying to encircle the enemy by forging a river downstream, hoping to catch the Chinese in a trap. My unit got aboard the tanks and we took off, arriving at a stream with the lead tank entering the water. A disaster happened immediately. The water depth was deeper than anticipated, causing the tank to become disabled, with the crew bailing out to escape drowning. Personal gear was removed from the tank with word that a tank retriever was on its way.

I waited to see if the mission would continue with the remaining tanks crossing in a different location. After a considerable length of time had elapsed, I was informed that my protection was no longer needed and I was ordered to return to my company. This was easier said than done. With the sun starting to set, I now had to retrace my route on foot to our original pickup area. I moved my unit, marching until I stumbled onto a column carrying supplies to our forward companies. I was relieved to know I was heading in the right direction. Finding a place in the column, we moved at a snail's pace with frequent stops to rest. It was dark, as we walked alongside a river, with stretches of high rocky banks extending almost to the water. It was a slow journey all night long with the sunlight slowly peeking in, making it easier to see where we were headed.

I finally found my company and reported to my platoon leader. He was glad to see that I had made it back safely and ordered me to take my section to an area he had selected. Seeing how tired and worn out we were from moving all night, he instructed me to just sack out and not waste time digging in. It seemed like I had just fallen asleep when the platoon leader sent a runner to get me. I walked over to where he rested, asking what he wanted. I was informed that tanks were being brought up and it would be my mission to supply protection for the tanks in their effort to open the road from our new location. I got my group together and walked over to where the tanks waited for us.

We got aboard the tanks and started moving down the narrow road. I could see our rifle company Marines occupying positions atop the ridges as we passed. Leaving our friendly area behind, we entered no man's land not knowing what to expect. It didn't take long to find out. The enemy was waiting, as the lead tank made a turn around a bend. That's when all hell broke loose. The machine gunner on the first tank opened fire and we scrambled off the tanks, making our way to the area around the bend. There we came across a group of Chinese soldiers with their hands up. In their possession was a 3.5 rocket launcher. One of our newest anti tank weapons developed to use against enemy armor. It must have been liberated in one of the successful victories the Chinese had had in North Korea. This gun almost succeeded in destroying one of our tanks. In his initial burst, the tank gunner had the right range, but was just a hair high. The slugs from his gun ricocheted from the Chink's head, making dents in his skull that looked like a ball peen hammer had worked it over. There were more soldiers originally with the group, but quite a few took off when we approached. Our captives said at least thirty more were with the group that had taken to the hills as we got close. The oddity about this capture was that none of the captives had any personal weapons on them.

We still had a mission to accomplish in clearing the road. Proceeding forward with our prisoners in tow, shortly we made contact with a Marine Recon unit that was clearing the road toward us. Information was exchanged and I asked them to relieve us of our prisoners. The

Recon unit wanted no part of them. They just wanted to be taken to where our activity had occurred. Detached from the tanks that were now closing in on Marines coming toward them, we backtracked and stopped at the bend, with the Recon Marines scaling the hill in search of more Chinese. I took the prisoners off the road and checked them over real well. These prisoners appeared to be well fed and may have been North Koreans instead of Chinese. One with a shitty grin on his face had a black pen he was waving with his hand. He figured we would take it from him, and he was right. There weren't any other items in their possession, so we got back on the road to rejoin our company. The other prisoners carried the Chinaman with dents on his head. Flies were now picking at his wounds.

Once back at my company CP, I asked where I should leave the prisoners. I was instructed to take them to a temporary stockade where more prisoners were being held. After being relieved of my prisoners, I stopped to chat with a fellow Marine who told me a rocket battery had moved into the area with a fire mission scheduled to take place shortly. I watched the prisoners as they sat on the ground relaxing, not knowing a rocket fire mission was about ready to blast off. I had moved from behind the rocket units and had gone a short distance when all hell broke loose. When the guns opened fire, the prisoners went crazy scurrying for cover and trying to find shelter from the dust and debris caused by the rocket propulsion and blown in their direction. They looked scared and bewildered, but at least they were still alive.

On return to my company, I was ordered to take my section to join a rifle company on the high ground located in the area where our prisoners were taken. It was dark when we got there and it started to rain. We rigged up shelter halves for protection. It was still raining in the morning when word came to move out. We moved atop the ridgeline until word came to descend to the road below. I used my poncho for protection again, hoping I could escape the soaking that I had experienced once before. The rain came down, causing the hill to become slippery. Some descending Marines lost their footing and now found themselves on the ground covered in mud as they slid down the slope.

Once on the road, we rejoined our company as it moved forward. There were tanks on the road moving with the column. The rain coming down gave me a cold, chilly feeling and my clothing was again getting wet, so I walked over to one of the tanks that had stopped, hoping the exhaust heat would warm me up. The heat felt good, but only while near the tank. Then the cold took over. Others had the same idea and shortly Marines near the tanks started to pass out. In trying to keep warm, they were put to sleep with carbon monoxide fumes.

Right Place, Right Time

After spending some time on the road with the tanks, the column started to move out. The rain finally stopped. My company moved into a valley, with each unit assigned a place to dig in for the night. My section had about settled in when I got word my unit was to relocate to Dog Company. I got my troops ready and, to my surprise, found a Dog Company guide waiting to take us in.

I reported to a platoon lieutenant. He was glad to see the additional firepower to anchor his defensive position. I asked where he wanted to place the machine guns. He led me to his defensive position and gave me the option on where to place them in the line. I looked over the area and told him I would have to take over some of his foxholes. I expected an argument, but he said, "No problem", and relocated those I selected, allowing some of my tired crew to move in without digging. The night was brutal with shells raining down, hitting the lower slopes or whistling as they flew over. No one got hit during the night and there was no attack. In the morning, the Lieutenant told me that I was free to rejoin my company. He again provided a guide to take us back down the ridge.

On return to my company area, I was shocked to see the complete devastation of the area where we were going to spend the night. It looked like a firing range had used it for practice. The heavy volume of shells going overhead during the night had landed in the location we had originally intended to stay. To say fate was kind to us that night would be an understatement.

I decided to change my socks that were still damp from the day before. I sat down, leaning against a wheel of a supply trailer. I had my shoes off, ready to slip the socks off, when an explosion occurring on the other side of the trailer rocked me with a dirt shower. Boy! That was too close for comfort. I looked around, saw a bunker, and dove in while more shells came raining down. In the bunker, I finished changing my socks, coming out when the shelling stopped.

Activity picked up. Since I had some prior tanker protection exposure, I was given another mission. I was told to find some tanks and report to the commanding officer for instruction. They couldn't tell me where they were located, but I was pointed in the general direction of where they were supposed to be. I was on my own again. It took some time but I finally found the tanks and reported to the officer in charge, awaiting his orders. It was now late in the afternoon and it seemed like a great deal of confusion was taking place. I was told to stay loose and wait for further instructions. The tank officer was waiting for word to move out on a mission. Finally, he came over to tell me that I wasn't needed any more and to report back to my company. This was another assignment resulting in no enemy confrontation.

The following morning, I was assigned to Dog Company for a company probe of enemy defensive positions with .81mm mortars and my section attached. We got started early. I could tell this was going to be a bad day. Enemy artillery was active right from the start. As we neared their lines shells were pouring down with a massive barrage, trying to take us out. The company C.O. instructed my section to stay back and set up a defensive perimeter around the .81mm mortar crew. I deployed my men and then went over to check in with the mortar crew section leader. Their guns were set up ready for a fire mission.

All this time, enemy artillery shells were seeking victims in the area. I found myself a spot to use for cover, and at the same time, I watched as the enemy shells were hitting and exploding. I noticed one shell had detonated about forty yards away. The next shell landed about twenty yards closer in the same line where I rested. Call it fate or just luck, but something urged me to go back and check to see if any of my men were hit. After making sure none were casualties, I returned to the mortar area and found the crew knocked out. I looked at the ground where I had sought shelter earlier. In its place was a newly created crater caused by the next shell that landed. Maybe that souvenir I picked up of Jesus had saved my life. I sure was lucky to cheat the grim reaper. With the mission complete, dead and wounded were evacuated and we returned to rejoin our company.

That night we found ourselves against a bank in a field. Toward evening, a tanker group pulled up alongside to spend the night. The additional firepower gave us a sense of security. We felt they offered excellent protection underneath, should the enemy shells come raining down, but one of the tankers convinced us otherwise. The next day, moving in a draw with bullets zinging overhead, one member of my unit by the name of Robinson, yelled, "Hey, Sarge, do you get a Purple Heart if you get shot in the fingers?" I replied, "You sure do." He then said, "Is it right that it takes two Purple Hearts to be taken out off combat?" I answered, "That's what I've been told, unless you volunteer for extended duty." After that exchange, I saw Robinson walking with his one arm outstretched over his head, hoping some enemy marksman would shoot one of his fingers off. He was willing to lose a finger or two in exchange for a more serious wound or being killed.

I was again assigned to a rifle company, only this time it was Easy Company. I was immediately impressed with the lieutenant in charge, who took what I considered an extra effort in his troop placement. He was a spirit of inspiration. All night long, he kept checking his people, voicing encouragement to keep them alert. With his constant vigil of foxhole positions, I doubt, whether he got any rest that night. Another first for me that night was exposure in the use of searchlights deflecting light off the clouds, resulting in better vision at night and thus depriving the enemy of sneaking up to our positions. There were also cameras set up at night, covering a fixed area of enemy terrain. They registered gun flashes and then overlaid on shots taken during daytime. It was a system used to locate and then take out enemy field guns. The Chinese made no attempt to probe our lines. In the morning, I again reported back to Weapons Company.

Souvenir Rifles

One day we came across an arsenal of Chinese weapons that were collected and placed in a pile. Word came down that anyone who wanted a souvenir rifle could choose one from the pile. The beauty of the offer was that any weapon properly tagged would find its way to the owner's home courtesy of Uncle Sam. I checked the pile, selecting a rifle that I claimed as my own. I wasn't the only Marine who wanted a memento to take home. I recall seeing rifle company Marines lugging Chinese weapons, as souvenirs, while climbing some of the rugged Korean mountains. It was a choice whether the prize was worth the effort. The extra weight was a deterrent factor in the decision making. In our case, since we did all our travel on foot with everyone about maxed out in the weight he carried, this offer was one that couldn't be refused.

I was looking forward to having my new trophy occupy a select place on a wall at home. It would be a reminder of my hardships in Korea. I never did receive my souvenir. Years later I learned that an Army general, I believe his name was Mitchell, confiscated all these weapons and sold them as a profit venture for his own benefit.

On the Move Again

Once again the companies were on the move, with the battalion committed to retaking ground held by the enemy. My section, marching on a country lane, watched as a battle took place on a high ridge located on the left side of the road. A rifle company was engaged in trying to displace the enemy from that defensive position. Another rifle company was in position to leapfrog and take the next objective once the first obstacle was taken. When the column made a temporary halt, I watched as wounded Marines made their way down the hill. A wounded buddy, who had suffered machine gun wounds to his legs, was being carried in a poncho down the slope. Just as they reached the road near where I rested, he let his bowels go, wet his pants, and died, before medics could tend to his wounds. It was heartbreaking to see another Marine give his life to make South Korea a free nation.

With the break over, my unit now approached a Korean house. There was an old mamasan standing outside to greet us. I was really surprised at her dress attire. She wore a skirt from the waist down that resembled a burlap bag, with the upper body bare. We entered the house. There were two females inside, one holding an infant child. The infant had open sores on his face and arms and there were flies buzzing and biting at the lesions. The other gal was a real beauty. It became evident that the good looking gal had a problem. She gave us the impression that she was mentally unstable, with the other girl trying to convince us that her companion was mentally gone. My men couldn't believe this good looking girl was loco.

Leaving the house, I was told to set up a machine gun on a lower section of a ridge overlooking open terrain, with word that we were there only temporarily. I left one squad in position in front of the house, while I took the other squad up the hill. Once the gun was set up, everyone took off their packs and relaxed. Shortly thereafter, I heard someone yell, "Quit throwing stones." The individual who made the remark was resting on the ground above me. I looked up, hoping to see what was happening. Once again I heard the same request, but saw no one playing games. In my section there was a Navajo Indian by the name of Wolf. He had gotten thirsty and needed a drink of water. He reached into his cartridge belt that was resting on the ground and pulled out his canteen. As he pulled it out, it didn't seem right. It felt empty. It also made a noise like a solid object was inside. Taking the cap off and turning it upside down, a rifle bullet dropped into his hand. What a beautiful way to pick up a souvenir and a story to go with it. Once again we were lucky that only a canteen was the lone victim to whoever was shooting at us.

During the night, I was assigned to man a roadblock covering the approach to a valley my company CP was located in. It was my responsibility to make certain no one planted any mines in the road leading to our position. It was a dark night with artillery shells whistling over our head. Looking up at the clouds, there were an unusual number of premature shells or "airbursts" exploding in the sky above. I heard someone say, "It's starting to rain," but I felt no drops as I wrapped a poncho over me. The shelling continued throughout the night, with airbursts still visible on a regular basis. It never did rain. However, as we were getting our equipment together to move out in the morning, one of my guys yelled, "Hey! Look what happened to my rifle." The stock on his weapon was shattered. The airbursts we observed during the night were the mistaken rain that fell. Once again, we managed to escape with no injuries. It was easier to replace the rifle stock than the person that carried it.

During the day, a truck loaded with ammunition arrived in the company CP area and ran over a land mine, with ammo flying all over the place. There was an ammo storage dump in the area, covered with tarps serving a dual purpose. In addition to ammo, bodies of dead Marines were also placed under the tarps, until evacuated to the rear. Curious individuals walking in the vicinity lifted the tarp to see if anyone they knew was among the dead. It wasn't a pleasant sight viewing the ghastly wounds on those that died.

We moved out the next day, coming to a clear field in front of us with a large hill roughly 500 yards to our right. A halt was called, so I sat down scanning the hill. In the distance I could see a Chinese unit marching toward us, then moving beyond the hill up ahead. When the break was over we started out, heading for the same ridge. Reaching the hill, we started to move around the base in single file. There appeared to be something lying just off the pathway that looked like a doll. As I got closer, I could see it wasn't a doll but a very small girl dressed in brightly colored red and white clothing. She appeared to be asleep. Although I couldn't see any physical injuries, the girl would never again enjoy the pleasures of playing and aging with others. Her rest was eternal. How she happened to be abandoned in such an isolated place was a mystery. There were no other bodies around.

We started to scale the ridge and dug in for the night without seeing any enemy soldiers. We spent the night on fifty percent alert, expecting the enemy I saw earlier to attack at night. There was a brief eruption of gun fire activity elsewhere on the ridge and then silence for the remainder of the night. The next morning word came down that a heavy machine gun unit from our company on a section of the ridge had been overrun during the night. I was told the gun watch got careless and fell asleep, resulting in other members of the unit bayoneted in their sleeping bags. The first thought that entered my mind on hearing this news was whether it was the same gun crew that was careless during our guerrilla chasing days. This was a real morale shocker. You just couldn't be careless if you wanted to survive. It must have been a patrol checking our perimeter and they succeeded in killing some of our sleeping Marines and then withdrawing. The next day we relocated to another ridge, digging in for the night. Enemy artillery shells kept landing in the area, causing casualties in the platoon. One Marine had his leg blown away, but my section escaped without harm.

Special Treats

Things were looking up. While in the hills, our ration supplies at times were augmented with a variety of candy, cigarettes and toilet articles like dental cream and shaving supplies. Candy and cigarettes were always acceptable, but the other goods were over abundant and buried in the ground. Every now and then, loaves of bread were included in the rations and shared equally among the men in the section. One day we had a surprise when with the bread issue came some oleo that served as a spread. I recall heating my share of oleo in my canteen cup and dipping what little bread I had into it. Man! That was a real treat.

Another time, word came down that doughnuts would be available with the meal ration. It was just one per man, but at least our Spartan days were improving. Chewing tobacco was scarce and in demand by those addicted to it. There was also an abundant supply of chewing gum with the special rations. With the daily field rations issued, one never knew what kind of menu came with it and usually resulted in buddy trades until the right exchange took place.

There was a rotation system the regiment used with battalions on line. Every so often, our unit was pulled off the line and placed in reserve. Everyone looked forward to a breather when it came. This last operation was a real trial for everyone. In reserve, the company C.O. gave us information on our participation in the last campaign against the enemy. At his debriefing meeting, he told us Weapons Company had sustained the most casualties in the battalion in our last engagement. This was just the opposite of the normal casualty reports because usually the rifle companies got chewed up worst. Another bit of information that was surprising, dealt with the good looking dummy. This girl wasn't loco at all, but turned out to be a North Korean agent. She just couldn't deceive her own people.

We were in a valley with the sun beating down. Everyone received a beer ration of two cans per man, but the suds were warm. Someone found a small stream not more than a yard wide running from a mountain. The water was cool so it became our fridge. It didn't take long before the cans cooled and the beer was a pleasure to drink.

The weather turned hot and sticky, with nights just as bad. Reserve didn't last long before word came to saddle up because we were going back on line. As I got ready to move out, I was told that a chaplain had arrived and would hold mass for those interested in attending. As a religious man I decided to go, hoping the Lord's blessing would shield me from harm's way in the near future. Mass was held in a wooded field with the Chaplain using a jeep as his altar, granting all in attendance absolution with communion given to all. It was surprising the large turnout attending mass, including the C.O. of Dog Company.

We were now moving along the road, when up ahead I saw what looked like vehicles off to the side of the road. Coming closer, I saw that there were Chinese soldiers seated on the back of one of the enemy trucks left behind by the enemy. The Air Force caught this truck in the open, killing the troops sitting in the back. The bodies were covered by a white mass of maggots having a feast. I had never seen anything like that before.

Vehicles and Horses

Off the road, I saw what appeared to be a small Marine supply trailer resting under a tree. Adjacent to the trailer on the ground were boxes of hidden enemy supplies. The immediate area was an arsenal of camouflaged ammunition and shells. We hit the jackpot. The ammunition supplies were so well hidden, it would have been impossible to spot them from a spotter plane. It was now our turn to savor the spoils of loot left behind by the enemy. The trailer, bearing Marine marking, must have been abandoned by Marines in North Korea and was finally liberated by its rightful owners. It still carried the 1st Marine Regimental markings. In addition to using their own vehicles, horses, and native porters, captured UN vehicles were also involved in moving enemy supplies from the Yalu to the combat zones.

Our troop movement took us through an area where a great number of dead horses were floating away. This was a new experience for me. I had never seen a wasted horse before, but what surprised me was how huge and long the entrails were spreading from the dead animals. With the amount of dead horses around, there had to be some that survived. Two members from my section came upon one, taking him as a prize courtesy of the enemy. Alabama and Stan were the captors. The horse had open saddle sores from what must have been heavy supplies lashed onto his back. Tender loving medical treatment by the new owners resulted in healing the sores. The burden of hauling shifted from Chinese to U.S. Marines. I don't believe the weight of two "A" frames was anywhere near the load that the Chinese had the horse carry. The horse wasn't a problem as yet and once we came off the line the fun began, with the owners allowing anyone in the section that wanted to try their horsemanship. Care and feeding responsibility rested with the owners.

Dear John

In the platoon, we had a youngster everyone called Mouse. He had acquired his name because of his small stature. At mail call one day, he received a Dear John letter from his girl back home. Once word got around, others in the platoon couldn't resist ribbing the daylights out of him. He was a broken man. Another incident that occurred involved a new replacement, a married man. Shortly after joining the section he went to sickbay, complaining of a medical problem. He came back refusing to believe he had a venereal infection, claiming loyalty to his bride back home. Then the games began with questions reverting to his wife's fidelity. Amazing how some individuals can dish it out, as long as, they aren't on the receiving end. All of the enlisted men had received a four page letter from "Ernie I Know" to "PFC Mac U Marine" warning us about the dangers of venereal diseases.

Innovations

Back in the hills again, a new policy was being tried in feeding the troops on line. Kitchen equipment normally never used with units on line was moved up and erected as close as possible to the rifle companies. Our positions were more stable now and we weren't on the move as much. Marines on line were now fed hot meals. A rotation system on a percentage basis allowed each man to share in this new experiment.

Another innovation dealt with hot showers. Trucks arrived in the battalion area to pick up Marines who came down off their hilltop positions on a percentage basis. They drove them to a portable shower unit in the rear where dirty uniforms were shed. Once showered down, Marines moved outside to find piles of clean clothing laid out on the ground. I don't know where the clothes came from, but each item had a name stenciled on it. A decision had to be made whether to accept a clean outfit with someone else's name or go back to wearing our own dirty dungarees. I went through the piles until I found recycled clothing to fit with a different name stenciled on each clean item I selected. I often wondered if Graves Registration made any errors as a result of misnamed, recycled clothes.

While waiting for transportation back to my company, a Marine approached me asking if I wanted my picture taken. He had a Polaroid camera and was willing to snap a photo of me for a mere price of $3.00. I asked what unit he belonged to. He told me he was assigned to a supply unit in the rear and figured he could make some money by selling instant photos to soldiers, who wanted to send home a current photo as a remembrance. His contact in Japan supplied him with all the film he needed. He was right. Marines paid his price just to see how rugged they looked with beards and long hair. I was a patsy and had a shot taken. No one in the unit had a camera. This was a new experience for me that someone would milk another Marine for profit while troops in the line lived a close camaraderie relationship with one another. Those who live in combat and rely on one another to survive can only experience this closeness. The camera bug must have been one of the capitalists the Chinese described in surrender leaflets found on the hills in our area. I doubt if he would have been as enterprising with a transfer to a rifle company.

Feeling the Heat

Many Marines discarded their steel helmets, until a major crackdown took place. Medical reports of increasing head injuries suffered by troops in the field, led to an all out effort to correct the problem. A scheduled inspection by company officers led those without steel tops scrambling to find replacements because there would be discipline for those without one. Maybe the British Marines didn't use them, but we had thankful Marines alive, because they had worn their life-saving helmet. Our first platoon leader was a prime example.

The weather conditions we operated in now got good and hot. This had a demoralizing effect on every one of us. Walking toward a hamlet, I found myself gagging from the stench that permeated the air in the immediate area. Koreans used human waste to fertilize their farmlands, a practice dating back to their ancestors. Most of the natives in the area wore masks over their face to protect themselves from the dust stirred up by gusty winds. Shelters to store human waste were called honey houses. When the land needed fertilizers, this waste was the answer.

I was now one of an alarming number of Marines that became casualties of dysentery. I reported to sickbay and was treated with sulfa drugs. More Marines were disabled due to dysentery than combat injuries. Steps to alert everyone to this new enemy required a medical doctor to hold a special informational lecture. He went on to explain just how easily this could happen. Within our field rations were plastic spoons that we usually saved after usage and placed in our dungaree pockets to be used when needed again. What none of us knew was the harmful exposure to germs that were in the air as a result of human fertilization the natives used in crop planting. The winds blew the dust onto our clothing or germs spread on our clothing by our digging and then living in foxholes. How did this medical doctor expect us to remain sanitary? We couldn't wash before eating, and our hands were always exposed to dirt that handled the food we ate.

One day word came down that a Marine had just dropped dead. Everyone questioned whether there was a new bug in the area and whether it was contagious. It turned out to be an isolated case with no more mysterious deaths involved. The recent germ warfare stories kept everyone alert.

Dropped from the Air

From our hill positions, on occasion, we watched as American planes flew overhead blaring a surrender message to Chinese soldiers below and dropping leaflets to use, as a surrender pass. The Chinese in turn had their response with surrender leaflets. In their propaganda leaflets, they appealed to the working man's loyalty. Why should he risk his life for the capitalists that were raking in the money, enjoying the safety of home while the GI was dying to make him richer? One night while dug in on the line, word came down to expect a night bombing raid from the enemy. This had never happened before. Extra precautions were taken should the information prove to be accurate. It turned out to be a false alert.

I was a witness to American airpower used against the enemy and procedures to protect friendly troops. Although I, myself, had never been strafed or bombed, a section leader in my platoon told me how he had been accidentally strafed by American planes, while assigned to a rifle company. He said it was a scary feeling and his unit was lucky to survive without incurring any casualties. Friendly fire could be deadly at times. A unit on line zeroed in the approaches to its positions with mortars and artillery fire. Observers from those supporting units assigned to a line company were responsible for registering their guns on these approaches. There were many incidents of premature shell explosions or short rounds, with shrapnel falling on friendly troops causing deaths and wounds. I had my own share of close encounters with short rounds, but managed to survive being hit. Others weren't as lucky.

Short Timer

Each month, more old timers were selected for rotation home. I missed a few and wondered if I would survive long enough to make one. Our platoon officers came and went. The new one in command was Lieutenant Meyers. I believe he may have been the fourth to assume command of our platoon. The platoon leader normally spent his time at the company CP and was hardly ever seen. The assignment to move the men out usually went to a section leader. Only on rare occasions would the platoon leader take the point. There was a time when word came to saddle up. The platoon leader was moving toward me and stopped to tell me I was to take the point and to move my section out in single file, without telling me where I was headed. When I raised the question, he looked at the open terrain leading toward a mass of hills. It was my turn again to test the land ahead.

I took my section and moved out into no man's land. With all the recent enemy activity, I expected the area to be loaded with land mines, never knowing if my next step would be my last. We had to rely on the old system of passing information verbally from one man to another down the line, until orders to halt reached me from the rear. It was another day of movement with no enemy contact to create any action.

We had to relocate again. It was late in the afternoon, when orders came to bed down for the night. The area we occupied was just forward of our artillery guns. Just as the sun started to set, we got a taste of what to expect throughout the night. The first salvo had everyone jumpy and trying to hide from the noise. The guns barked all night long with no one able to get any shuteye. It was a blessing when morning came and we moved out. There were a lot of tired people on the move that day, and when a break was called the troops just slid to the ground trying to catch a few quick zzz's.

We continued moving along a river, passing a dead Marine that was pulled out of the stream. He was a black Marine with his body badly swollen. It raised a question as to whether he had gone swimming and drowned or if there was another reason for what had happened. That evening we occupied a ridge and started digging in for the night. I hadn't completed my foxhole when a flurry of activity took place. Someone spotted some movement coming toward our position and hand grenades started to fly. It was an enemy patrol checking the line, but the grenades prevented any further action.

We still had the horse with us, but not for long. It wasn't certain if the enemy spotted the horse first or the troops on the move. The heavy volume of enemy shells raining down sealed his fate. The horse was turned over to South Korean troops and may have been used as horsemeat for their table.

Special Assignment

On June 19, 1951, the Lieutenant came over to give me a special assignment. I was to lead a party of Korean supply workers loaded with mortar ammo badly needed by one of our mortar sections, attached to a rifle company. I got my section together, found the ammo party ready, and we took off. The closer we got to where we were headed, the volume of enemy shelling increased. The shells landed to my left, right, front, or rear. It was a touch and go situation. I kept checking my ammo carriers to make certain they were still with us. I didn't want any to break and run. We had to keep moving or sooner or later we were bound to become casualties. The Lord was with us, protecting each and everyone, as we finally found the unit in need.

Our mission was completed, but the shelling didn't stop. A member of my section by the name of Brown became a casualty when a shell exploded near him. Shrapnel turned his face blood red. I couldn't tell how badly he had been hit. A corpsman looked to his needs. It was like being in hell with shells exploding close, seeking to claim a victim or victims. Finally, I moved my section away from our company mortars to an area where it wasn't as bad and then proceeded to return to where I started. I aged a great deal that day and was thankful that I had survived (although just barely) another day.

There were rumors circulating concerning cease fire negotiations. That's all they were, rumors. We kept running into heavier enemy resistance. Artillery and mortar fire kept increasing daily. We had moved into an area near a river with relief scheduled to take place the next day. Early the following morning while still dark, a relief column drove up close to the line with replacements to affect the move. They came under heavy enemy mortar fire immediately, taking a large amount of casualties. This change of targets took some pressure away from our position, but not for long before they again shifted their guns and peppered our position.

We got the word of what happened to our relief. The manpower loss wouldn't have been as great, if the trucks had dropped the replacements off further to the rear, and then marched them forward to our positions. The Chinese were on the other side of the river in excellent position to hear or see any activity taking place on the road. Although it was still dark when they arrived, it was the noise that got the convoy in trouble.

Word finally came to move out. Leaving the area with shells still raining down, we came to a ridge where heavy activity was taking place. The hill was enveloped in a cloud of smoke as a rifle company was battling the Chinese for control of the high ground. It's hard to explain, but every now and then, we entered an area having a terrible putrid odor that lingered following heavy shelling. It had the stench of death associated with it.

We were into July with another group of Marines going home. The platoon sergeant was on the list in the next draft, with speculation as to who would fill his vacancy. Since there were no NCOs higher then sergeant, there was a guessing game as to who would be given that opening. The suspense ended when a staff sergeant arrived with the new replacement group and filled the position.

Crawling with Ants

We were involved in a campaign called the Punch Bowl, with relocation on a regular basis. I was assigned to a rifle company with word that the whole division would be coming off the line shortly. It was dark as I moved my unit below a ridge line with orders to dig in. Enemy shells were falling, looking for a target. I broke out my shovel, dug my hole, and tried to make myself comfortable. Shells were still seeking victims, as I crawled into my sleeping bag. Shortly, I could feel something crawling all over me. I tried not to panic as a colony of large ants invaded my sleeping bag, causing me great discomfort. My hair, face, and body inside my clothing were covered with ants. I could feel them crawling all over me, leaving me no choice but to vacate my foxhole, pull all my gear out, and shed my clothing to try to cleanse all the ants from my body and clothes. In the dark, I couldn't see how badly my other possessions were infested. I didn't know if I should dig another foxhole or take my chances in the open. Enemy shells were pounding the ridge to the right and left, with an occasional round in my vicinity. There might have been more ant hills around, so I decided not to take the chance of spooking them. I decided to forgo relocating and spent the night above ground but close to the hole just in case there was a need to use it as a last resort.

At the break of dawn, I could see all the red ants scooting around. This was a horrible experience for me. I was the only one with bad luck that night, and now had to eliminate all the pests I could find from my sleeping bag and clothing The other guys in my section couldn't believe my nightmare, checking out my foxhole to verify my night in hell. With this episode behind, word came to saddle up and move to an area behind a ridge away from where the shells were hitting. Finally the magical orders came saying our unit was being relieved. Shortly thereafter, transportation arrived and we were trucked to a rest area within the Punch Bowl. This became our home for the next few weeks.

On the Rotation List

Our stay here was different from our previous rest locations. The last time I had seen a movie was prior to my departure from the States, but here there was an outdoor area set up to show movies. One night while at the movies, the picture playing was boring so some of us decided to leave. We weren't very far from the front lines and could see gun flashes in the sky from activity taking place where the combatants were at war. I was leaning against a tree looking at the sky, when I spotted a large orange object moving in the direction toward where the flashes were occurring. My companions also saw the object. It looked like a large cigar. We continued to watch it as it disappeared from sight, questioning: "What the hell was that?"

On August the , a memorial service was held to honor 201 Marines from the 7th Regiment that perished in combat from January until now. A program listing activities for the ceremony was handed out and included names of each Marine that died. However, the list lacked the individual unit each deceased served in, depriving one of knowing which company lost the most manpower.

One day, the whole battalion was assembled to receive a speech from a chaplain. He told us that we had the foulest, dirtiest, filthiest mouths in the world, using profanity in our everyday conversations that wasn't acceptable. He asked every Marine to kindly make an honest effort to clean up his vocabulary. That speech was a shocker to some, but also put smiles on the guilty parties.

I finally received the news I was awaiting. The eighth rotation draft for September had my name on it and all I had to do was mark time until my departure. That would give me a year of service in Korea and maybe a chance to cheat the grim reaper. Some Marines in heavy machine guns felt this was an ideal time to brew some raisin jack. As I walked by the brew master's cot, I could hear the rumbling taking place in five gallon water cans that were used for making the brew. The officer in command of this platoon was a Naval Academy graduate. He sat with his men, as they tasted the fermented juices, with the men knowing there would be no excuses in performing their job assignments the following day, especially after spending the night drinking with their platoon lieutenant.

One night, I was invited to a tent, where Japanese liquor purchased from a ship manned by Japanese sailors was available. It was a sociable drinking party, until the liquor was used up. It was still early to call it a night, so a corpsman in the tent volunteered to free up some 190 alcohol from the medical supply tent. On his return, he came back with the goods, plus cans of grapefruit juice to mix with the alcohol. When all the booze was gone, everyone returned to his tent. A patrol was on the agenda the following morning. I woke up with a miserable hangover. The exercise of walking in the hills didn't help in relieving my pain. I suffered throughout the patrol, feeling like I was going to die. I just drank too much of the poison and now I was paying the price. I vowed that once I felt better, I would never do it again.

My unit missed seeing Bob Hope, when he entertained the troops in Korea, but while at the Punch Bowl, we had the luxury of being entertained by Jack Benny and Errol Flynn. There were some girls with them. One was Rita Gam and a girl called Pat, whose last name I can't recall. This was a real morale boost for everyone. A few days later, a country western group came down to entertain us. Uncle Sam was doing all he could to keep our morale high.

While in reserve, word came down that the military script now in our possession would be changed, giving everyone a chance to convert to the new currency. There were many Koreans attached to our unit who held a sizeable amount of the old currency. Would they lose all their savings? Some had worked hard to acquire their money and didn't want to lose what they had. The issue was raised with their superior officers. I didn't understand why every so often they kept changing this currency, but I heard someone say it was related to black marketing.

One day, the regimental commander held a troop formation for Marines being decorated. After the medals were awarded, he gave a speech in regards to rumors of cease fire negotiations. Colonel Nickerson, our new regimental commander, told us not to be misled by the communists. He said, "You can't trust those people. If they succeed in winning a cease fire in Korea, then you can look forward to a confrontation in Indo China." Was he a prophet alerting us to another war?

Alvin Brown, who was wounded on June 19th during the worst enemy mortar barrage thrown at us, reported in after recovering from his facial wound. We all questioned him on his injury, wanting to know how badly he was hit. Brown told us his face was splattered with teeny metal slivers, but he was lucky not to have been hit in the eyes. He showed us how he went about picking the metal fragments out of his face, once a blackhead appeared. It was good to have him back.

Toward the end of August, rumors started to circulate that our reserve status was about to change. Shortly thereafter, the rumor became a reality and all the troops got ready to return to combat. The big question to be answered dealt with all Marines that had been notified about going home in the September rotation. Would they accompany their units into battle or would they be left behind? A decision was made that all rotation Marines would remain with their companies until notified to detach. We only had a few more days left to go before going home, and now I had to face the grim reaper again. It was now a matter of time to see who would win the encounter.

Trucks came and drove us to our assembly area a short distance behind the front lines. Marines were now relieving troops in the line, while my battalion was held in reserve. What a stroke of luck. I would gain a few additional days before being committed. While in this reserve area, I was exposed to a new esprit de corps adopted by the rifle companies. Some Marines were planting state flags adjacent to their foxholes. It looked like Flag Day with all the colors unfurled and whipping in the wind. Finally, word came to move up and we took our position on line.

Enemy shelling was our main menace, until the day I was told by my platoon leader to report to an assembly area, located in a valley at the bottom of the hill we occupied. There was a bridge over a stream used by vehicles to cross. During the night, an ammunition truck crossing the bridge suffered a direct hit from enemy shells.

Leaving My Unit

On the morning of my departure, I had to control my feelings now that I was leaving the unit that I had led, since the beginning of the year. We had shared some good times and bad times together, plus many hazardous experiences with no one attached to my unit making the KIA list. I felt damn good about that record, but felt bad about those that were wounded. Saying goodbye was hard, as I wished everyone in my section luck in their stay in Korea with a word of encouragement that they, too, would survive until rotated home.

I high tailed it down the hill, crossed the bridge, and sat down where other Marines were also waiting for transportation to take us back to the rear. While waiting for trucks to come and take us out, an explosion went off less than a hundred yards away on the other side of the stream. That got everyone's attention, especially since we were going home and needed just a little more time before we were out of harm's way. It turned out that a rotation Marine walking to join us stepped on a land mine. His return home would be in a body bag. He was one of many identified to leave in the September draft that was either killed or wounded. Someone had made a decision to keep all rotation troops within the companies, when the Division went back on line. Some may have written letters to their families alerting them that they were coming home, only to have the misfortune to become casualties of war.

Finally, the trucks arrived. We got aboard and were on our way climbing a steep mountain road that took us out of the valley where we were still in artillery range. Once we reached the top of the pass and started down the other side, everyone relaxed. Some expressed their feelings about leaving the place. One of the men going home pulled out a bunch of wallets and started searching through them. Another man asked him where he had gotten all the billfolds. He said he got them from dead Marines. They had no more use for them, so he helped himself to their possessions. I heard about ghouls like him preying on the dead, but never expected to see one sitting next to me crowing how stealing from his own buddies was better for him than for someone in Graves Registration.

The trucks took us to an airfield, where we boarded a plane bound for Pusan. On landing at the airstrip, I was surprised to see a mobile Red Cross canteen vehicle waiting to greet us. They were serving coffee, doughnuts and soda pop to troops arriving for rotation home and to new replacements waiting for transportation to fill our vacancies.

While waiting for transportation to our staging area, we shared our time with the new replacements that were curious to know what awaited them and were asking us all sorts of questions. Finally, trucks came and took us to our destination. We arrived at a fenced in area where tents were used to house incoming and outgoing personnel. This was our home while waiting for our ship to arrive. For some reason it was late, but it was expected to make port in a day or two.

Almost immediately in the tent area, information was given each Marine regarding the high venereal disease rate in Pusan and Japan, with literature issued to make us aware of the many different strains of V.D. that servicemen had picked up from contact with local prostitutes. It was a good policy to make everyone aware how a little fun could wreck their lives, if they should become infected. I had seen some bad VD cases aboard the aircraft carrier that required complete isolation during their healing period. It wasn't worth the price to pay for a little pleasure.

Liberty was available to anyone who wanted to visit the city. I went into town and ended up at the American Red Cross canteen, where all kinds of activities were available for use by servicemen. Traditional coffee and doughnuts flowed freely. Baked cakes and other food were also plentiful, with various games to keep us entertained. I heard some horror stories about the Red Cross from World War Two vets, who had unpleasant dealings with them, but all my encounters were just the opposite.

The following morning, scuttlebutt had it that a prostitute had been smuggled through the fence during the night and was housed inside one of the tents. I was surprised to learn she was only ten years old. War sure corrupts people.

Onboard the McKinley

The USS McKinley (AP-114) finally arrived. It was September 11th and we were now aboard ship waiting to sail. An Army band arrived to see us off with the last song: "So Long, It's Been So Good To Have Known You" echoing as the ship weighed anchor. They couldn't have chosen a more appropriate tune as a tribute to see us depart. As I stood on deck leaning over the rail watching Pusan disappear, I could honestly say:

I'm going home, my Lord,
I'm going home.
To forget my misery
Caused by rain and snow.
To recall days spent
In warmth and cold.
To forget shot and shell
Pain and death.
How has the year
Aged my growth?
Tears stream my cheeks
With thoughts of friends
Young and old.
Buried in graves
That no one knows.
I find myself confused.
Why was I
One of the chosen few?
Spared to live
While others died.
Only the Lord knows.

The next stop on the way home was Kobe, Japan. We arrived in port on September the 12th, where the sea bags that we left aboard the Bayfield were stored. At the warehouse, I rummaged through the baggage, until I found my sea bag. Checking the contents of the bag, I found hardly anything inside. It would take a complete refit to replace what I would need when I was back stateside.

Back aboard ship, liberty was available immediately, giving everyone a chance to hit the bars or do some shopping for bargain items. Liberty was a preview of what I had missed for over a year and awaited me once I got back to the good old USA. On September the 15th, we were back at sea. It was ironic that exactly a year ago on that very day, I had arrived in Kobe on my way to war, and today I was on my way home.

There were all kinds of work details assigned to rotation Marines aboard ship. Holding the rank of sergeant was a benefit that excluded me from any work assignment. I became a man of leisure just resting and hitting the mess line when chow was served.

One day, I met a mortar man from my company who came aboard in Kobe. We got to rehash our time in Korea. He told me some weird stories. His mortar section was with Fox Company at Toktong pass where some Chinese prisoners were taken in their attempt to dislodge or annihilate all the Marines on that ridge. The situation became critical. It finally came down to where no one could be spared to guard them. A decision was made to execute all the captives, before the Chinese made any more attacks on the hill. He was given the assignment to terminate one of them. I asked him how he went about doing it. He told me he took his captive aside, gave him a cigarette, walked behind him, and then when the prisoner wasn't looking, took out his pistol and shot him in the head. I then asked him how he felt after completion of his assignment. With so many dead and wounded Marines in that encirclement, he said they had no other choice.

He played poker with anyone that had money and during his stay in Korea had been able to send home $5,600 that he won gambling. He may have been lucky in cards, but he sure lost out when it came to women. He was recuperating in a hospital in Japan when his passion ran away with him. He played with fire and got burned. His souvenir for sleeping with prostitutes was a bad case of Japanese warts. I felt sorry for him. He was a married man with a V.D. problem that wasn't cured and that he was taking home to his family.

The last night before arriving stateside, I met a Marine on deck who knew I had a North Korean burp gun I was taking home as a souvenir. He asked me if I would be willing to swap it for a Chinese rifle that he was taking home. Since it was against the law to bring home automatic weapons, I asked to see his rifle. We then made the switch. This was the only trophy I brought back from my days in hell.

Pizza Pie

On September the 26th, we arrived at San Francisco, California, docking at Treasure Island with a band to greet us. This was our home, while undergoing physicals and processing to our next duty station. The person involved in processing asked what duty station was my preference. I was coming home with Don Ivers, a Marine from Ballston Spa, New York, who told me about a naval supply depot at Scotia, New York. He said Marines were stationed on this base and if asked for a choice of stations to ask for Scotia. I took his advice and we both were assigned there. He said he would take care of me on my arrival, because he lived just a few miles away.

While waiting for transfer orders, we had open liberty every day. Don and I went to see San Francisco and he took me to a restaurant for dinner. While browsing through the menu, Don said to forget the menu and just order a pizza pie. I asked him what a pizza pie was. He said, "You'll see once it gets here." He ordered two whole pizza pies with all the trimmings, one for me, and one for him. I had never seen, heard, or tasted a pizza pie until that night and what they brought me must have been a jumbo size that was way too large for one person to eat. After a couple of slices, I had my fill and then watched as Don devoured his whole pie. For a little man, he sure could eat.

Speaking of food, the first day back I went to eat chow at the mess hall, I entered the line, picked up my silverware, and proceeded through the serving station. When I saw what looked like breaded chicken, I asked for an additional portion. At the table I sank my teeth into what I thought was chicken, but it didn't taste like chicken. I made a comment on how different this chicken tasted. Everyone laughed at me saying I was eating rabbit, not chicken. That ruined my appetite. I had never eaten rabbit before. I decided that a hamburger in town would taste better.

Travel Misfortunes

Outside our barracks, one could look out into the bay and see Alcatraz, home to many hardened criminals. What surprised me the

119

most was how cool and damp the weather was here. With processing all but done, everyone was given a 30-day leave, with additional time for travel to one's next duty station. There were airline travel agents booking fares on charters to different parts of the country. This expedited getting home quicker. For a hundred dollars, I booked a flight out of Oakland leaving for New York City, where I would take a train to Buffalo from Grand Central Station. I expected to reach home in a day.

We left Oakland before midnight on October 3rd and landed in Amarillo, Texas, with a lengthy delay for plane maintenance. From there, we flew to St. Louis, Missouri, where the plane was again checked for engine problems. Our next stop was O'Hare in Chicago, where we had to deplane for six hours while necessary repairs were done to the aircraft. Once aboard and ready for takeoff, another delay occurred and we spent three more hours on the ground waiting until finally the lame duck got off the ground. In the air and on our way to New York, we hit a rainstorm that rerouted our flight to Newark, New Jersey. Now a bus was needed to take us to New York.

With all the misfortune encountered up to now, I wondered what else could go wrong. It wasn't long before an embarrassing incident took place. While seated on the bus waiting for the driver to come aboard, a man in a blue suit entered the bus and someone yelled, "Hey driver, when the hell are we going to move?" The man in the blue suit turned around to answer the question, "I'm sorry, but I'm not the driver." The man had a pair of gold bars pinned on his shoulders and was a captain dressed in a new blue U.S. Air Force uniform. It was an honest mistake made by someone who had never seen a blue Air Force uniform before.

It was well past midnight, when we finally reached Grand Central Station. It had taken more than 20 plus hours to fly from California to New York. I should have been home by then, but now I had to wait a few more hours for my train to leave. I checked my baggage and decided to visit a tavern for a couple of beers. The place I hit was crowded. As I made my way to the bar, I heard a loud voice say, "Hey Marine. Come over here." An elderly patron called me over and introduced himself. He looked me over, called the bartender, and told him to cover all my drinks on the house for the night. He also had some money in his hand that he shoved into my coat pocket. I asked him why he was doing that; he said that it was his way of showing his appreciation for the great job that Marines were doing in Korea. What a way to start a leave. I finally returned to the station and boarded my train when it arrived. I was just hours from being home.

War Behind Me

We pulled into Central Terminal in Buffalo, where I hailed a cab, and within minutes, I was home. The first person I saw was my Dad, who was working in the garden. When he saw me exit the cab, he came over and gave me a warm welcome, then helped me with my baggage. It was the first week of October and now I could forget the war.

While at home, I was curious as to what had become of my two friends that entered the service with me. I learned that John P. was with the 1st Marines, when they hit the beach at Inchon. He sustained a shoulder wound during the Seoul operation and since his return stateside had gotten married. My other friend Hank B. was still stateside and got himself married to one of the girls we met at the wrestling match, while on leave from boot camp. At least, we were all still alive.

It was amazing how quickly the days flew and my leave ended. I reported to Scotia on November 14th, expecting to see Don Ivers and looking forward to the good times he promised me. It just didn't pan out. Don's orders were changed while on leave, directing him to report to Camp Lejeune instead. There were many Korean veterans assigned to this base. I came across one from my unit, who I was under the impression had been wounded in our first enemy confrontation. I asked him how badly he was injured. He said he wasn't wounded at all. What actually happened to him was he suffered a heart attack. I couldn't believe a young, healthy, 18 year old could have a heart attack, but he said he got so scared that he did have one.

One day, I was told to see the commanding officer, a Major Nourse, who told me a formation was to be held to present me with a medal.

121

The base commanding officer, a Navy captain, was to do the honors. The award I received was the Bronze Star medal with Combat V for my activity on June 19, 1951 for leading a mortar laden supply unit to its destination, while under a tremendous enemy mortar barrage. A photograph taken that day appears in my memoir photo album.

Another time, there was a solicitation for blood donors going around, so I volunteered to donate. There were Marines on base alive that day, because of blood transfusions they had received after being wounded. One Marine told me he received nine pints of blood when he was wounded. This donation was to take place in Schenectady on the 10th of December. Transportation would be made available for everyone who signed to give. While waiting for transportation on the designated day, I was singled out with another sergeant to ride into town with the Commanding Officer of the supply depot. Upon arrival at the Red Cross receiving station, we were ushered immediately into a room and became the first to have our blood drawn. Local news photographers took our pictures while we were lying on tables giving blood. Our chauffeur, the base Captain, then drove us back to base. The next day our picture, with a write-up, was carried in the local newspaper back home.

On December 17th, I received a promotion to staff sergeant. I had just had my sergeant chevrons sewn onto all my newly issued clothing, now I had to replace all the old chevrons with new ones. I found the Scotia base was a Mecca for anyone who had undergone a starvation diet. The daily menu was beyond expectation. The food was out of this world. When steaks were served they covered the whole plate. Quarts of milk were a standard staple at every meal. Frog legs also made the menu, with weekend meals served to order. It almost made one want to re-enlist again, just to be fed in this grand style.

I couldn't get over how friendly the people were in that area. There were many civilian employees working on the base, and during the holidays such as Thanksgiving, Christmas and New Year's, invitations were received from families asking Marines or sailors to share their holiday meals with them. There were many eager volunteers to fill the requests.

I applied for and was granted Christmas leave. This was my first Christmas at home with my family since my enlistment and I had a great time. My time in the Corps was winding down. Just a few more days and it would just be memories with experiences money can't buy. The hour finally arrived when I had to sign all sorts of papers and then I was given my discharge. I was a civilian again. The war in Korea was still going hot and heavy, but I had served my time in Hell and had survived to tell about it.

Post-Military

After being discharged from the Marine Corps, I went to work for Ford Motor Corporation in an apprenticeship program for four years as a tool and die maker, a skilled tradesman. In 1952, I met Emily Ruszczyk at a carnival in Lackawanna. We began dating and married a year later in 1953. We had five children: two daughters and three sons. On September 15, 1954, Barbara Ann was born. Rosemary was born on April 28, 1957. Richard J. Janca was born June 27, 1958. David William was born July 14, 1961 and Jamie Christopher was born February 3, 1973.

After working for Ford Motor Company, I transferred to General Motors, where I became a Union Representative for 23 years. I worked 33 years for General Motors. I became active in many veterans' organizations, including the Chosin Few, 1st Marine Division, Conrad Kania Marine Corps League, Military Order of the Purple Heart No. 187, DAV No. 135, Korean War Veterans and the Matthew Glab Post No. 1477.

Awards Received

For my service in Korea I earned the following awards:

- Purple Heart WIA 27th Sept 1950
- Purple Heart Certificate WIA 27th Sept 1950
- Presidential Unit Citation(s) with ribbon & 2 Bronze stars awards 1stMarDiv, Reinforced, for service in Korea Sep-Oct & Nov-Dec 1950, & 1951
- Bronze Star With Combat "V" for action on 19th June 1951
- Korean Service Medal
- United Nations Service Medal
- Korean Presidential Unit Citation(s) awarded to 1st Mar Div for service in Korea 1950 and 1950-1953
- Stars Authorized For Korean Service Medal
  - North Korean Aggression - 27th June 1950 to 2 Nov 1950
  - Communist China Aggression - 3 Nov to 24 Jan 1951
  - First U.N. Counter Offensive - 25th Jan to 21 Apr 1951
  - Communist China Spring Offensive - 22 Apr to 8 Jul 1951
  - U. N. Summer-Fall Offensive - 9 Jul to 27 Nov1951
- There is also a DIVISION MEMORANDUM issued to all Chosin Marines and the 41st Royal Marine Commandos by Oliver Smith, Major General, USMC, Commanding General 1st Marine Division.

Epilogue

Richard was a religious man and felt lucky his life was spared in the war. He was able to marry, provide for his family and enjoy a simple life. He had a great sense of humor and enjoyed making a good deal. Richard's gift of gab molded his character. He had many phrases and traits that became "classics" with his family. Richard was charitable to many causes and believed what goes around comes around. He wanted his children to find their own way, without interference from him. He was proud of all of them and was called "Popster" by his grandchildren.

His good luck charm was the war relic he found. Strikingly, the day the photo was found, Dale Erickson, a fellow Marine, stood by Richard's side. They entered an area called Massacre Valley. Rain had doused most of the fires from the battle. The ground and surroundings were wet except for the photo. It was marred by only a few drops of water and had singe marks from the fires that had smoldered.

The picture that Richard found and carried in his wallet until his death was called, "**The Boy Jesus**" and was originally painted by Florence A. Kroger. Florence was born in 1897 and died in 1980. Florence's work appeared on the front cover of the 1933 Home Magazine. Her images of Christ appeared in Catholic magazines such as the Cincinnati based Messenger. Florence became a successful commercial artist who lived in California in the 1950's. Aside from the religious works of art, she was most remembered for her nursery images in soft pastels. Florence's art work was quite extensive and she grew up as part of a famous family, the Kroger's of Cincinnati, Ohio. The Kroger's are known for the line of grocery stores they started which operate today in many parts of the country.

At Richard's funeral, Monsignor Burkhart eulogized the strong faith Richard had and conveyed the story of the "Boy Jesus" photo. It turns out that the Monsignor had two stories to tell about the picture. Growing up, the Monsignor was also familiar with the photo, since a similar picture hung on the wall in his family home. Call it Coincidence or Divine Intervention? Let's hope we all find the answers in eternal life.

Obituary - Richard Janca

Richard Antoni Janca, of Orchard Park, New York, decorated Marine veteran and longtime United Auto Workers representative, died Monday, July 27, 2009. Burial was held Thursday (July 30) at Our Lady of Victory Basilica with entombment at Hillcrest Cemetery.

The son of the late Joseph and Julia Zynczak Janca, Richard married Emily Ruszczyk in 1953. She survives. He is the father of Barbara (William) Rose, Rose (late Paul) Stewart, Rick (Elaine) Janca, David (Lynn) Janca and Jay (Jessica) Janca. He is the grandfather of Lisa, David, Derek, Ricky, Gregory, Dillon, Alexa, and Kyle Janca, and the late Angela Sliwinski. He is brother of Michael (Florence) Janca and the late Edward (Florence) Janca. He is also survived by many nieces, nephews, relatives and friends.

Richard was a Marine veteran of the Korean War. He was awarded the Bronze Star and Purple Heart for his service to his country. He was in many pivotal campaigns during the war, including the Chosin Reservoir, where U.S. forces, although greatly outnumbered, fought their way out of being surrounded by the Chinese and Korean forces.

Richard was employed at Harrison Radiator Division of General Motors for 33 years. For 25 years he represented the United Auto Workers skilled trade union at his plant. He was active in many organizations, including the Chosin Few, 1st Marine Division, Conrad Kania Marine Corps League, Military Order of the Purple Heart No. 187, DAV No. 135, Korean War Veterans and the Matthew Glab Post No. 1477.

Memorials may be made to OLV Homes of Charity or Father Baker Renovation Fund.

The Messenger

Messenger Art Collection, one of the most comprehensive and eclectic collections in the US, celebrates its 100th birthday in 2013. The collection had been languishing in obscurity and disrepair until it was acquired by a new owner, Albert Babbitt, in June 2010. The 5,000+ works of art in the collection are now safely ensconced in an environmentally sophisticated environment, and many of its treasures have been archivally restored.

Mr. Babbitt has formulated a plan for the collection which honors Frank Messenger and his personal vision, bringing the art into the "light" once again and making it available for scholars, museums, historians, designers, and art collectors through its engaging website: www.messengerartcollection.com .

"We are honored to have been approached by Barbara Rose to identify and share this moving oil painting by Florence Kroger, *"The Boy Jesus"* which helped inspire the author's beloved father to survive the challenges of war," said Mr. Babbitt, "as Ms. Rose communicates very vividly the timeless message of hope, commitment, patriotism and profound courage.

www.ingramcontent.com/pod-product-compliance
Lightning Source LLC
Chambersburg PA
CBHW060313050426
42448CB00009B/1820